Condensed Matter Physics: Advanced Principles and Applications

Condensed Matter Physics: Advanced Principles and Applications

Edited by Jaron Finley

CLANRYE
INTERNATIONAL
www.clanryeinternational.com

Clanrye International,
750 Third Avenue, 9th Floor,
New York, NY 10017, USA

ISBN: 978-1-64726-603-5

Cataloging-in-publication Data

Condensed matter physics : advanced principles and applications / edited by Jaron Finley.
p. cm.
Includes bibliographical references and index.
ISBN 978-1-64726-603-5
1. Condensed matter. 2. Matter. 3. Solid state physics. 4. Physics. I. Finley, Jaron.
QC173.454 .A38 2023
530.41--dc23

For information on all Clanrye International publications visit our website at www.clanryeinternational.com

Contents

Preface

Every book is a source of knowledge and this one is no exception. The idea that led to the conceptualization of this book was the fact that the world is advancing rapidly; which makes it crucial to document the progress in every field. I am aware that a lot of data is already available, yet, there is a lot more to learn. Hence, I accepted the responsibility of editing this book and contributing my knowledge to the community.

Condensed matter physics refers to the branch of physics which studies the microscopic and macroscopic physical properties of matter. It particularly deals with the solid and liquid phases that originate from electromagnetic forces among atoms. The Bose–Einstein condensate originating in ultracold atomic systems, superconducting phase revealed through certain materials at low temperatures, and the antiferromagnetic and ferromagnetic phases of spins on crystal lattices of atoms are some of the exotic condensed phases. The magnetic, elastic, optical, thermal and electrical properties of liquid and solid substances are also studied in condensed-matter physics. Its study comprises the principles of electromagnetism, quantum mechanics and statistical mechanics. There are various applications of condensed matter physics in developing devices such as solid state laser, liquid crystal display and optical fiber. This book includes some of the vital pieces of work being conducted across the world on condensed matter physics. It aims to serve as a resource guide for students and experts alike and contribute to the growth of the discipline.

While editing this book, I had multiple visions for it. Then I finally narrowed down to make every chapter a sole standing text explaining a particular topic, so that they can be used independently. However, the umbrella subject sinews them into a common theme. This makes the book a unique platform of knowledge.

I would like to give the major credit of this book to the experts from every corner of the world, who took the time to share their expertise with us. Also, I owe the completion of this book to the never-ending support of my family, who supported me throughout the project.

Editor

1

Superconducting Josephson-Based Metamaterials for Quantum-Limited Parametric Amplification

Luca Fasolo, Angelo Greco and Emanuele Enrico

Abstract

In the last few years, several groups have proposed and developed their own platforms demonstrating quantum-limited linear parametric amplification, with evident applications in quantum information and computation, electrical and optical metrology, radio astronomy, and basic physics concerning axion detection. Here, we propose a short review on the physics behind parametric amplification via metamaterials composed by coplanar waveguides embedding several Josephson junctions. We present and compare different schemes that exploit the nonlinearity of the Josephson current-phase relation to mix the so-called signal, idler, and pump tones. The chapter then presents and compares three different theoretical models, developed in the last few years, to predict the dynamics of these nonlinear systems in the particular case of a 4-wave mixing process and under the degenerate undepleted pump assumption. We will demonstrate that, under the same assumption, all the results are comparable in terms of amplification of the output fields.

Keywords: superconductivity, metamaterial, Josephson effect, parametric amplification, microwave photonics

1. Introduction

In the last decade, microwave quantum electronics received a substantial boost by the advancements in superconducting circuits and dilution refrigerators technologies. These platforms allow experiments to be easily carried out in the mK regime, where the detection and manipulation of signals in the range $3 - 12$ GHz reaches energy sensitivities comparable to a single photon [1].

Solid state microwave quantum electronics is founded on a building block that has no analogous in quantum optics: the Josephson junction [2]. This, in fact, is a unique nondissipative and nonlinear component that represents the key element of a large series of quantum experiments.

Furthermore, microwave quantum electronics allows the exploration of the so-called ultrastrong coupling regime [3], hard to be reached in quantum optics, and it is worth mentioning that nonlinear resonator can be exploited to access relativistic quantum effects and quantum vacuum effects. To give an example, the Lamb shift [4] effect has been observed in superconducting artificial atom [5], while the dynamical Casimir effect [6, 7] has been promoted by properly engineered superconducting waveguide [8].

From the very beginning, superconducting electronics has been pushed by the strong interest coming from the quantum computation and information community. However, it has been only recently shown that a new concept of 1D metamaterial with embedded several Josephson junctions enables strong photon-photon on-chip interactions [9], allowing experimentalists to engineer dispersion relations that drive the waves traveling along artificial waveguides [10, 11]. These concepts and technologies allow the control and tunability of the wave mixing process. As an example, a weak signal traveling in a metamaterial can interact with a strong pump tone at a different frequency, activating the so-called parametric amplification [12]. The class of devices where these phenomena are promoted is commonly known as traveling-wave Josephson parametric amplifiers (TWJPA) and represents the solid state analogous to optical χ^n nonlinear crystals [13].

It has been shown that TWJPAs can act as quantum parametric amplifiers by reaching the so-called quantum limit [14]. With the purpose of a comparison to the state-of-the-art commercially available l ow-noise am pl ifiers, these latter can operate at $\omega/2\pi = 4$ GHz adding $k_B T_n/\hbar\omega \approx 10$ noise photons having a noise temperature of $T_n = 2$ K, while Josephson-based amplifiers can reduce this added noise up to 1/2 photon, or even 0, depending on its working configuration.

The capability to beat the quantum limit is related to the so-called phase-sensitive amplification process, where the metamaterial can operate in degenerate mode (degenerate parametric amplifier, DPA), acting on two waves (signal and idler) at the same frequency ($\omega_s = \omega_i$) by amplifying and de-amplifying their position and momentum quadratures, respectively. In this view, DPA enables the preparation of squeezed states in the microwave regime. Even in the nondegenerate mode (nondegenerate parametric amplifier, NDPA, i.e., $\omega_s \neq \omega_i$), the phase-preserving nature of the quantum parametric amplification results in the entanglement condition among the signal and idler generated photons, composing a two-mode squeezed state [15]. It is worth mentioning how such a quantum state is an example of Einstein-Podolsky-Rosen state [16], where correlations between signal and idler are stronger than that allowed by classical theory [17].

It should be evident how superconducting electronics not only has demonstrated to be an ideal platform for microwave quantum parametric amplification but also has pushed forward the research field focusing on the generation of nonclassical radiation with attractive potential applications in metrology and quantum informa-tion processing.

2. Historical evolution of the traveling-wave parametric amplifiers

The theory of a new concept of microwave amplifier was developed by Cullen [12] in 1959. In his paper, Cullen showed a novel mechanism of periodic transfer of power between a pump tone and a signal traveling in a transmission line composed of a voltage dependent capacitance per unit length. A nonlinear component of an RLC circuit can change periodically the resonance frequency of the whole system, leading to a novel way of making broadband amplification, the so-called parametric amplification. In **Figure 1**, we report two toy models for parametric amplification in mechanical systems with their electrical counterparts.

One of the first realizations of Cullen's idea was made by Mavaddat et al. in 1962 [19]. The signal line was basically a low pass filter, in which the shunt elements were similar varactor diodes. There, the nonlinearity was given by the specific capacitance-voltage relation of the varactor diodes, which is highly nonl inear for

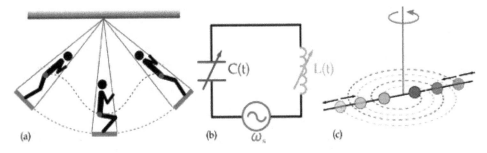

Figure 1.
(a) Sketch of a swing process. An oscillating system at a frequency ω_s is excited by parametric amplification via periodical changes of the center of mass position at a frequency $\omega_p = 2\omega_s$. (b) LC circuit with variable (nonlinear) C and L components. The case in which the capacitance C is periodically changed in time is the circuit analogous to the mechanical system represented in (a), while the case having an oscillating inductance L mimics the condition sketched in (c), consisting in a torque pendulum with variable inertia momentum [18].

relative small voltage values. In this pioneering experiment, a gain of 10 dB and a bandwidth of 3 MHz were shown.

After the theorizing and the subsequent discovery of the Josephson effect [2], it was understood that an easy way to embed a nonlinear component into a transmis-sion line and simultaneously reduce losses was to build a nonlinear inductance made of superconducting material, exploiting a Josephson junction as a source of nonlinearity following the vanguard idea by Sweeny and Mahler [13]. There, the parametric amplifier was modeled by a first-order small-signal theory with the same approach adopted to predict the behavior of GaAsFET transmission line amplifiers. The proposed design consisted of a superconducting thin-film niobium transmission line, composed by a coplanar waveguide integrating a large number of Josephson junctions.

The first realization of a traveling-wave parametric amplifier embedding a series of Josephson junctions was possible due to the PARTS process developed at IBM [9]. Exploiting niobium/aluminum technology, Yurke et al. [20] reported the construc-tion and characterization of a coplanar waveguide, in which the central trace was composed by an array of 1000 Josephson junctions. The experiment was there performed in reflection mode, by terminating one end of the device with a short, l eading to a relative high gain of 16 dB but a narrow bandwidth of 125 MHz and a noise temperature of 0.5 ± 0.1 K. The mismatch between the theoretical model and the experimental data has resulted in the understanding of a lack of a complete description of the physics behind this device when working in a small-signal regime. The study of the collective behavior of groups of Josephson junctions forming a transmission line has been an active field of study of several theoretical works [21, 22]. Subsequently, the use of numerical analysis [23] helped in clarifying how wave propagation acts inside this kind of transmission line, giving information on cutoff propagation, dispersive behavior, and shock-wave formation. An analyt-ical model of a Josephson traveling-wave amplifier of greater complexity was developed by Yaakobi et al. [24]. There, a transmission line made of a series of capacitively shunted Josephson junctions was considered.

One of the main limitations concerning the maximum achievable gain, common to all the TWJPAs concepts, is represented by the phase mismatch between the different tones into the line. In particular, it has to be noticed that even though the incoming waves can be in phase, photon-photon interactions between different tones (cross-phase modulation) or the same tone (self-phase modulation) lead to a modification of the phase of the traveling tones themselves. Indeed, quantum mechanical ly speaking, the power transport between the pump and the signal waves takes place through a photon energy conversion between the pump and the signal. This means that for an efficient energy exchange, conservation of both energy and momentum needs to take place. The latter condition is the corpuscular analogous to the phase matching requirement between the different electromagnetic waves. An engineering solution to overcome this problem is represented by the so-called resonant phase matching (RPM) [11]. O'Brien et al. analyzed this method theoretically on a simple transmission line made of a series of Josephson junctions capacitively shunted to ground operating in the so-called 4-wave mixing (4WM) regime. In their model, they shunted the transmission line with several LC resonators with a resonance frequency slightly above the pump tone. Doing this, they were able to show the rise of a stop band in the dispersion relation, which is able to re-phase the pump with the signal tones by changing the pump wave vector, favoring the wave mixing.

O'Brien's design was realized not long after [10] using Al technology. In their design, the unit cell of the transmission line was composed by three single nonlinear Josephson cells, the shunt capacitor was made using low-loss amorphous silicon dielectric and a resonator was placed after each group of 17 unit cells. The device showed a maximum gain of 12 dB over a 4 GHz bandwidth centered on $\approx$ 5 GHz. Moreover, the authors explain that variations of 2–3 dB in the gain most likely come from imperfect impedance matching between sections and at the level of the bond pads.

A similar design was adopted by Macklin et al. [14] to prove experimentally the capability of a TWJPA combined with the RPM technique to be used as a reliable tool for qubits readout. In this paper, the TWJPA, based on Nb technology and a different RPM periodicity, was first characterized, showing a gain of 20 dB over a 3 GHz bandwidth. Moreover, the quantum efficiency of the amplifier was tested when coupled with a 3D transmon qubit, leading to an efficiency value of 0.49 ± 0.01. A key point of this experiment was the proof that a single TWJPA could be able to perform the readout of more than 20 qubits, thanks to its high dynamic range and multiplexing capabilities. RPM has shown remarkable capabilities and is a promising technique to overcome phase mismatch. It can be implemented in multiple ways [25], by the way, it has to be noticed that this method requires an increase of design complexity, lower tolerances on the constructing parameters, and longer propagation lengths (2 cm–1 m).

Another option to solve the mismatch problem was suggested by Bell and Samlov [26], who proposed a self-phase matching transmission line embedding a series of asymmetric superconducting quantum interference devices (SQUIDs). The remarkable feature of this design is that it does not need any resonant circuit to achieve phase matching. This TWJPA is indeed able to tune the nonlinearity of its SQUIDs just through the use of an external magnetic field. Zhang et al. realized that this design [27] proves the wide tunability on positive and negative values of the Kerr nonlinearity by a magnetic flux and its capability to assist phase matching in the 4WM process. The 4WM process is intrinsically affected by phase mismatch because it takes origin from a cubic (Kerr-like) nonlinearity of the current-phase relation of the SQUIDs composing the TWJPA, getting unwanted effects from self-phase and cross-phase modulations.

Zorin showed [28] that by embedding a chain of rf-SQUIDs into a coplanar waveguide, it is possible to tune both the second and third order nonlinearities of their phase-current relation. This is a totally a novel approach to the TWJPA, since the possibility to use a quadratic term as a source of nonlinearity, allows to work in the 3-Wave Mixing regime (3WM), as theorized by Cullen 57 years before. It is well-known that 3WM has several advantages when compared to 4WM. Firstly, it allows to operate with a minimal phase mismatch. Secondly, it requires a lesser pump power to achieve the same amplification per unit length. Eventually, it separates signal and idler from pump tones, easing the engineering of the experimental setup by removing the requirement of heavy filtering in the middle of the amplification band. A proof of principle based on the Zorin's layout [29] showed a gain reaching 11 dB over a 3 GHz bandwidth.

A step forward in controlling the metamaterial nonlinearities was attempted by Miano et al. [30] achieving an independent tune of both second and third order terms in the current-phase relation by adjusting the bias current in some inductive circuits surrounding the transmission line. This technology takes the name of symmetric traveling-wave parametric amplifier (STWPA), its peculiarity arising from the symmetric arrangement of the rf-SQUIDs that compose the transmission line. This device concept represents the state-of-the-art in the field, allowing the exploration a wide portion of the control parameters space, leading to a maximum estimated gain of 17 dB and a 4 GHz bandwidth.

3. Theoretical models for a 4WM process in a TWJPA

In the last decade, different theoretical models have been developed to predict the behavior of an electric transmission line containing an array of Josephson junctions, employed as nonlinear elements. In this section, we will focus on those models developed to predict the behavior of a TWJPA in the particular case of a 4WM process, under undepleted degenerate pump approximation (i.e., assuming that the power held by the pump wave is at first approximation constant and larger than the one owned by the signal and the idler). We will firstly focus on the classical theory proposed by Yaakobi et al. in 2013 [24] and O'Brien et al. in 2014 [11], in which the behavior of the transmission line is derived imposing the current conservation in the system. This starting assumption leads to the definition of a partial differential nonlinear equation that can be turned into a system of coupled mode equations, providing the expression of the amplitude of the pump, signal, and idler tones along the transmission line. Subsequently, we will discuss two different quantum approaches for the description of the parametric amplifier dynamics. The first one, proposed by Grimsmo and Blais in 2017 [31], exploits a Hamiltonian based on continuous-mode operators to derive, in an interaction picture frame, a device's output field. The second one instead, proposed by van der Reep in 2019 [32], derives a system of coupled mode equations for the creation and annihilation quantum operators starting from a Hamiltonian based on discrete-mode operators.

The theories presented in this chapter will be based on a series of simplifying assumptions, whose experimental realization could be difficult to be obtained. For instance, in a real device, the undepleted pump approximation is hardly respected along the entire extension of the device because, along the line, the pump tone transfers a non-negligible amount of energy to the signal and idler one. The depletion effects, resulting in a reduction of the gain and of the dynamics-range of the amplifier, have been studied both in a classical and quantum frame [28, 33].

In all of these models, a lossless electrical circuit composed by the repetition of an elementary cell, whose structure is shown in **Figure 2**, is taken into account. In order to standardize the notations, we assume that the Josephson junctions embedded in the transmission line are identical (i.e., they have the same critical current I_c) and that the current flowing through the n-th junction can be expressed through the nonlinear relation

$$I_{J,n} = I_c \sin(\varphi_n) \qquad (1)$$

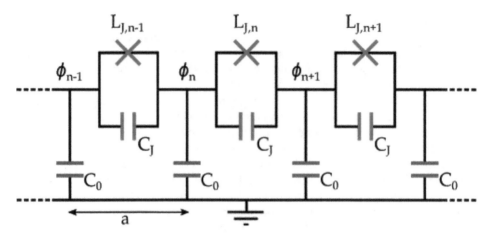

Figure 2.
Electrical equivalent representation of a repetition of Josephson junctions embedded in a transmission line. The junctions are modeled as an LC resonant circuit. The length of the unitary cell length is represented by a.

where $\varphi_n t()$ is the phase difference across the junction of the macroscopic wave functions of the two superconductive electrodes. The relation between φ_n and the voltage-difference across the junction is given by the Faraday's induction law

$$\Delta V_n = V_{n+1} - V_n = \frac{\Phi_0}{2\pi} \frac{d\varphi_n}{dt} = \frac{\Phi_0}{2\pi} \frac{d}{dt} \left[\phi_{n+1} - \phi_n \right] \tag{2}$$

where $\Phi_0 = h/(2e)$ is the magnetic quantum flux (with h the Planck constant and e the elementary charge), whereas $\phi_n(t)$ is the absolute phase in the n-th node of the circuit. The phase at the n-th node (ϕ_n) can be converted into a flux at the n-th node, and vice versa, through the relation $\Phi_n = (\Phi_0/2\pi)\phi_n$.

Furthermore, we define C_J the capacitance associated to the n-th Josephson junction and $L_{J,n}$ its inductance, defined as

$$L_{J,n} = \frac{\Delta V_n}{dI_{J,n}/dt} = \frac{\Phi_0}{2\pi} \frac{1}{I_c \cos(\varphi_n)} = \frac{L_{J_0}}{\cos(\varphi_n)} \tag{3}$$

where $L_{J_0} = \Phi_0/(2\pi I_c)$ is the inductance of the Josephson junction for a phase difference $\varphi_n = 0$.

The energy stored in the n-th Josephson junction can be expressed, using the definitions given in Eqs. (1) and (2), as

$$U_{J,n} = \int_{t_0}^{t} VI \ dt' = \int_{t_0}^{t} \frac{\Phi_0}{2\pi} \frac{d\varphi_n(t')}{dt'} \ I_c \sin(\varphi_n) \ dt' = I_c \frac{\Phi_0}{2\pi} \left[1 - \cos(\varphi_n(t)) \right], \tag{4}$$

under the assumption that $\varphi_n(t_0) = 0$, and approximated through a first-order power expansion as

$$U_{J,n} = I_c \frac{\Phi_0}{2\pi} \left[1 - \cos(\varphi_n) \right] = \frac{1}{2L_{J_0}} \Delta\Phi_n^2 - \frac{1}{24 L_{J_0}} \left(\frac{2\pi}{\Phi_0} \right)^2 \Delta\Phi_n^4 + O(\Delta\Phi_n^6) \tag{5}$$

where $\Delta\Phi_n = \Phi_{n+1} - \Phi_n$.

Finally, we assume identical coupling capacitances C_0 between the transmission line and ground.

3.1 The classical theoretical model

In this subsection, we will present the main steps for the derivation of the classical model presented in [11, 24]. Under proper assumption, this model allows to determine analytically the amplitude of the signal's and idler's waves along the transmission line.

Expressing the current flowing through each branch of the circuit presented in **Figure 2** in terms of absolute phases ϕ_n and imposing the current conservation in the n-th node (i.e., $I_{J,n-1} + I_{C_J,n-1} = I_{J,n} + I_{C_J,n} + I_{C_0,n}$), a differential equation for the absolute phase, in the case of a weak nonlinearity, can be obtained:

$$-C_0 \frac{d^2}{dt^2}[\phi_n] = -C_J \frac{d^2}{dt^2}[\phi_{n+1} + \phi_{n-1} - 2\phi_n] - \frac{1}{L_{J_0}}[\phi_{n+1} + \phi_{n-1} - 2\phi_n]$$
$$+ \frac{\Phi_0}{2\pi} \frac{1}{6I_c^2 L_{J_0}^3}\left[(\phi_{n+1} - \phi_n)^3 - (\phi_n - \phi_{n-1})^3\right] \quad (6)$$

where the last term derives from the first-order approximation of the nonlinear behavior of the Josephson's inductance.

Assuming the length a of the elementary cell is much smaller than the wave lengths of the propagating waves λ (i.e., $a / \lambda \ll 1$), the discrete index n can be replaced by a continuous position x along the line (i.e., $\phi_n(t) \to \phi(x, t)$) and the phase differences can be expressed, at the second order approximation, as:

$$\phi_{n+1} - \phi_n \approx a \frac{\partial \phi}{\partial x} + \frac{1}{2} a^2 \frac{\partial^2 \phi}{\partial x^2} \quad (7)$$

$$\phi_n - \phi_{n-1} \approx a \frac{\partial \phi}{\partial x} - \frac{1}{2} a^2 \frac{\partial^2 \phi}{\partial x^2} \quad (8)$$

In this way, it is possible to define a nonlinear differential equation for the continuous absolute phase $\phi(x, t)$:

$$C_0 \frac{\partial^2 \phi}{\partial t^2} - \frac{a^2}{L_{J_0}} \frac{\partial^2 \phi}{\partial x^2} - C_J a^2 \frac{\partial^4 \phi}{\partial x^2 \partial t^2} = -\frac{a^4}{2 I_c^2 L_{J_0}^3} \frac{\partial^2 \phi}{\partial x^2}\left(\frac{\partial \phi}{\partial x}\right)^2 \quad (9)$$

In the case of a weakly nonlinear medium, the dispersion law can be derived from Eq. (9), considering the left-hand side being equal to zero and imposing a plane-wave solution $\phi(x, t) \propto e^{i(kx - \omega t)}$:

$$k(\omega) = \frac{\omega \sqrt{L_{J_0} C_0}}{a\sqrt{1 - L_{J_0} C_J \omega^2}} \quad (10)$$

The solutions of Eq. (9) can be expressed, as shown by O'Brien et al. [11], in the form of a superposition of three waves (pump, signal, and idler) whose amplitudes are complex functions of the position along the line:

$$\phi(x, t) = \sum_{n=p,s,i} \text{Re}\left[A_n(x) e^{i(k_n x - \omega_n t)}\right] = \frac{1}{2}\sum_{n=s,i,p}\left[A_n(x) e^{i(k_n x - \omega_n t)} + c.c\right] \quad (11)$$

The case of a 4WM process with a degenerate pump can be taken into account by imposing the frequency matching condition $2\omega_p = \omega_s + \omega_i$. Replacing this particular solution in Eq. (9) and assuming that, along the line, the amplitudes are slowly varying (i.e., $|\partial^2 A_n/\partial x^2| \ll k_n |\partial A_n/\partial x| \ll k_n^2 |A_n|$) and that $|A_s|^2$ and $|A_i|^2$ are negligible (i.e., $|A_{s,i}| \ll |A_p|$, strong pump approximation), we obtain a system of three coupled differential equations for the amplitudes $A_n(x)$ that describe the energy exchange between the three waves along the line:

$$\frac{\partial A_p}{\partial x} = i\vartheta_p |A_p|^2 A_p + 2iX_p A_p^* A_s A_i e^{i\Delta kx} \tag{12}$$

$$\frac{\partial A_{s(i)}}{\partial x} = i\vartheta_{s(i)} |A_p|^2 A_{s(i)} + iX_{s(i)} A_p^2 A_{i(s)} e^{i\Delta kx} \tag{13}$$

where $\Delta k = 2k_p - k_s - k_i$ is the chromatic dispersion. The term ϑ_p is responsible for the self-phase modulation of the pump tone, while $\vartheta_{s(i)}$ is responsible for the cross-phase modulation between the pump tone and the signal or idler, respectively. These terms can be expressed as

$$\vartheta_p = \frac{a^4 k_p^5}{16 C_0 I_c^2 L_{J_0}^3 \omega_p^2} \quad \text{and} \quad \vartheta_{s(i)} = \frac{a^4 k_p^2 k_{s(i)}^3}{8 C_0 I_c^2 L_{J_0}^3 \omega_{s(i)}^2} \tag{14}$$

while the coupling constants X_n, depending on the circuit parameters, are defined as

$$X_p = \frac{a^4 k_p^2 k_s k_i (k_p - \Delta k)}{16 C_0 I_c^2 L_{J_0}^3 \omega_p^2} \quad \text{and} \quad X_{s(i)} = \frac{a^4 k_p^2 k_s k_i (k_{s(i)} + \Delta k)}{16 C_0 I_c^2 L_{J_0}^3 \omega_{s(i)}^2} \tag{15}$$

Expressing the complex amplitudes $A_n(x)$ in a the co-rotating frame

$$A_n(x) = A_{n_0} e^{i\vartheta_n |A_{p_0}|^2 x} \tag{16}$$

it can be demonstrated that, working under the undepleted pump approximation $|A_p(x)| = A_{p_0} \gg |A_{s(i)}(x)|$, the amplitude of the signal can be expressed as

$$A_s(x) = \left[A_{s_0} \left(\cosh(g_1 x) - \frac{i \Psi_1}{2g_1} \sinh(g_1 x) \right) + i \frac{X_{s(i)} |A_{p_0}|^2}{g_1} A_{s(i)_0}^* \sinh(g_1 x) \right] e^{i\frac{\Psi_1}{2}x} \tag{17}$$

where $\Psi_1 = \Delta k + (2\vartheta_p - \vartheta_s - \vartheta_i)|A_{p_0}|^2 = \Delta k + \vartheta |A_{p_0}|^2$ is the total phase mismatch and g_1 is the exponential complex gain factor, defined as

$$g_1 = \sqrt{X_s X_i^* |A_{p_0}|^4 - \left(\frac{\Psi_1}{2}\right)^2} \tag{18}$$

The total gain of an amplifier, composed by the repetition of N elementary cells, can then be expressed as $G_s(aN) = |A_s(aN)/A_{s_0}|^2$.

3.2 Quantum Hamiltonian model based on continuous-mode operators

A standard method to treat quantum superconducting circuits is represented by the lumped element approach [34]. In this latter, the Hamiltonian of the quantum circuit is straightforwardly derived from its classical counterpart by promoting fields to operators and properly imposing commutating relations. In this view, one can proceed by deriving the Lagrangian of a TWJPA composed by the repetition of N unitary cells, under first nonlinear order approximation, as

$$\mathcal{L} = \sum_{n=0}^{N-1} \left[\frac{C_0}{2} \left(\frac{\partial \Phi_n}{\partial t} \right)^2 + \frac{C_J}{2} \left(\frac{\partial \Delta \Phi_n}{\partial t} \right)^2 - E_{J_0} \left(1 - \cos \left(\frac{2\pi}{\Phi_0} \Delta \Phi_n \right) \right) \right] \approx \quad (19)$$

$$\approx \sum_{n=0}^{N-1} \left[\frac{C_0}{2} \left(\frac{\partial \Phi_n}{\partial t} \right)^2 + \frac{C_J}{2} \left(\frac{\partial \Delta \Phi_n}{\partial t} \right)^2 - \frac{1}{2L_{J_0}} \Delta \Phi_n^2 - \frac{1}{24 L_{J_0}} \left(\frac{2\pi}{\Phi_0} \right)^2 \Delta \Phi_n^4 \right] \quad (20)$$

where $E_{J_0} = I_c \Phi_0 / 2\pi = I_c L_{J_0}$. Under the assumption that $a/\lambda \ll 1$ it is possible, as performed in the previous subsection, to replace the discrete index n with a continuous position x along the line (i.e., $\Phi_n(t) \to \Phi(x,t)$) and approximate, at the first order, $\Delta \Phi_n \to a \, \partial \Phi(x,t)/\partial x$. Furthermore, extending the system via two lossless semi-infinite transmission lines (characterized by a constant distributed capacitance c_0 and a constant distributed inductance l_0), the Lagrangian can be expressed through a space integral extending from $x = -\infty$ to $x = +\infty$ as

$$\mathcal{L} \left[\Phi, \frac{\partial \Phi}{\partial t} \right] = \frac{1}{2} \int_{-\infty}^{\infty} \left[c(x) \left(\frac{\partial \Phi}{\partial t} \right)^2 + \frac{1}{\omega_J^2(x) l(x)} \left(\frac{\partial^2 \Phi}{\partial x \partial t} \right) - \frac{1}{l(x)} \left(\frac{\partial \Phi}{\partial x} \right)^2 + \gamma(x) \left(\frac{\partial \Phi}{\partial x} \right)^4 \right] dx$$

$$(21)$$

where $c(x)$ and $l(x)$ are the distributed capacitance and inductance of the system, defined as

$$c(x) = \begin{cases} c_0 & x < 0 \\ C_J/a & 0 < x < z \\ c_0 & x > z \end{cases} \quad \text{and} \quad l(x) = \begin{cases} l_0 & x < 0 \\ L_{J_0}/a & 0 < x < z \\ l_0 & x > z \end{cases} \quad (22)$$

$$\omega_J(x) = \begin{cases} \infty & x < 0 \\ 1/\sqrt{L_{J_0} C_J} & 0 < x < z \\ \infty & x > z \end{cases} \quad \text{and} \quad \gamma(x) = \begin{cases} 0 & x < 0 \\ (a^3 E_{J_0}/12)(2\pi/\Phi_0)^4 & 0 < x < z \\ 0 & x > z \end{cases}$$

$$(23)$$

where z is the length of the TWJPA, $\omega_J(x)$ is the junction's plasma frequency, and $\gamma(x)$ is the term deriving from the nonlinearity of the junctions.

From Eq. (21), one can easily derive the Euler-Lagrange equation whose form, for $0 < x < z$, is equal to Eq. (9), giving the same dispersion relation (Eq. (10)) under analogous assumptions. Instead, outside the nonlinear region, the wave vector turns out to be $k_\omega(x) = \sqrt{c_0 l_0} \omega^2$.

The Hamiltonian of the system can be derived from the Lagrangian by taking into account $\pi(x,t) = \delta \mathcal{L}/\delta[\partial \Phi/\partial t]$, the canonical momentum of the flux $\Phi(x,t)$:

$$H[\Phi, \pi] = \int_{-\infty}^{\infty} \left[\pi \frac{\partial \Phi}{\partial t} \right] dx - \mathcal{L}$$

$$= \frac{1}{2} \int_{-\infty}^{\infty} \left[c(x) \left(\frac{\partial \Phi}{\partial \Phi t} \right)^2 + \frac{1}{l(x)} \left(\frac{\partial \Phi}{\partial x} \right)^2 + \frac{1}{\omega_p^2(x) l(x)} \left(\frac{\partial^2 \Phi}{\partial x \partial t} \right)^2 \right] dx - \frac{\gamma}{2} \int_0^z \left[\left(\frac{\partial \Phi}{\partial x} \right)^4 \right] dx$$

$$= H_0 + H_1$$

$$(24)$$

where the term H_0 represents the linear contributions to the energy of the system, while H_1 is the first-order nonlinear contribution.

This Hamiltonian can be converted to its quantum form promoting the field $\Phi(x,t)$ to the quantum operator $\hat{\Phi}(x,t)$.

In direct analogy with Eq. (11), one can express the flux operator in terms of continuous - mode functions [34], such as $\hat{H}_0$ is diagonal in the plane-waves unperturbed modes decomposition:

$$\hat{\Phi}(x,t) = \sum_{\nu=L,R}\int_0^\infty \left[\sqrt{\frac{\hbar l(x)}{4\pi k_\omega(x)}}\hat{a}_{\nu\omega}e^{\,i(\pm k_\omega(x)x-\omega t)} + \text{H.c.}\right]d\omega \qquad (25)$$

where the subscript R denotes a progressive wave, while L denotes a regressive wave (i.e., $\hat{a}_{R\omega}$ represents the annihilation operator of a right-moving field of frequency ω). In [35], it is demonstrated that by replacing the definition given in Eq. (25) into the linear Hamiltonian, $\hat{H}_0$ takes the form

$$\hat{H}_0 = \sum_{\nu=R,L}\int_0^\infty \left[\hbar\omega\hat{a}_{\nu\omega}^\dagger\hat{a}_{\nu\omega}\right]d\omega \qquad (26)$$

(where the zero - point energy, which does not influence the dynamics of the amplifier, has been omitted).

Using the expansion of $\hat{\Phi}(x,t)$ introduced above, under the hypothesis of a strong right-moving classical pump centered in ω_p, and that the fields $\hat{a}_{\nu\omega}$ are small except for frequencies closed to the pump frequency (i.e., replacing $\hat{a}_{\nu\omega}$ with $\hat{a}_{\nu\omega} + b(\omega)$, where $b(\omega)$ is a complex valued function centered in ω_p), the nonlinear Hamiltonian $\hat{H}_1$ can be expressed under strong pump approximation, at the first order in $b(\omega)$, as the sum of three different contributions:

$$\hat{H}_1 = \hat{H}_{CPM} + \hat{H}_{SQ} + H_{SPM} \qquad (27)$$

In the expressions of these contributions, the fast rotating terms and the highly phase mismatched left-moving field have been neglected:

$$\hat{H}_{CPM} = -\frac{\hbar}{2\pi}\int_0^\infty d\omega_s d\omega_i d\Omega_p d\Omega_{p'}\sqrt{k_{\omega_s}k_{\omega_i}}\beta^*(\Omega_p)\beta(\Omega_{p'})\Upsilon(\omega_s,\omega_i,\Omega_p,\Omega_{p'})\hat{a}_{R\omega_s}^\dagger\hat{a}_{R\omega_i} + \text{H.c}$$

$$(28)$$

describes the cross-phase modulation,

$$\hat{H}_{SQ} = -\frac{\hbar}{4\pi}\int_0^\infty d\omega_s d\omega_i d\Omega_p d\Omega_{p'}\sqrt{k_{\omega_s}k_{\omega_i}}\beta(\Omega_p)\beta(\Omega_{p'})\Upsilon(\omega_s,\Omega_p,\omega_i,\Omega_{p'})\hat{a}_{R\omega_s}^\dagger\hat{a}_{R\omega_i}^\dagger + \text{H.c.}$$

$$(29)$$

describes the broadband squeezing, and

$$H_{SPM} = -\frac{\hbar}{4\pi}\int_0^\infty d\omega_s d\omega_i d\Omega_p d\Omega_{p'}\sqrt{k_{\omega_s}k_{\omega_i}}\beta^*(\Omega_p)\beta(\Omega_{p'})\Upsilon(\omega_s,\omega_i,\Omega_p,\Omega_{p'})b^*(\omega_s)b(\omega_i) + \text{H.c.}$$

$$(30)$$

describes the self-phase modulation. $\beta(\Omega)$ is the dimensionless pump amplitude, proportional to the ratio between the pump current $I_J(\Omega_p)$ and the critical current of the junctions I_c:

$$\beta(\Omega_p) = \frac{I_J(\Omega_p)}{4I_c} \qquad (31)$$

The function $\Upsilon(\omega_1, \omega_2, \omega_3, \omega_4)$ is the phase matching function, defined as

$$\Upsilon(\omega_1, \omega_2, \omega_3, \omega_4) = \int_0^z e^{-i\left(k_{\omega_1}(x) - k_{\omega_2}(x) + k_{\omega_3}(x) - k_{\omega_4}(x)\right)} dx \qquad (32)$$

Assuming the nonlinear Hamiltonian $\hat{H}_1$ as a perturbative term of the Hamiltonian $\hat{H}_0$ for which the continuous modes are noninteracting, and assuming the initial time of the interaction $t_0 = -\infty$ and the final time $t_1 = +\infty$, it is possible to relate the input field of the system to the output one introducing the asymptotic output field

$$\hat{a}_{R\omega}^{out} = \hat{U}\hat{a}_{R\omega}\hat{U}^\dagger, \qquad (33)$$

where $\hat{U}$ is the asymptotic unitary evolution operator (approximated to the first order in $\hat{H}_1$)

$$\hat{U} \equiv \hat{U}(-\infty, \infty) = e^{-\frac{i}{\hbar}\hat{K}_1} \quad \text{where} \quad \hat{K}_1 = \hat{K}_{CPM} + \hat{K}_{SQ} + K_{SPM} \qquad (34)$$

Working in the monochromatic degenerate pump limit $b(\Omega_{p'}) = b(\Omega_p) \rightarrow b(\Omega_p)\delta(\Omega_p - \omega_p) \equiv b_p$ (where ω_p is the pump frequency), the propagators take the form

$$\hat{K}_{CPM} = -2\hbar|\beta_p|^2 z \int_0^\infty k_{\omega_s} \hat{a}_{R\omega_s}^\dagger \hat{a}_{R\omega_s} d\omega_s \qquad (35)$$

$$\hat{K}_{SQ} = -\frac{i\hbar}{2}|\beta_p|^2 \int_0^\infty \sqrt{k_{\omega_s} k_{\omega_i}} \left[\frac{1}{\Delta k}\left(e^{i\Delta kz} - 1\right)\right] \hat{a}_{R\omega_s}^\dagger \hat{a}_{R\omega_i}^\dagger d\omega_s + \text{H.c.} \qquad (36)$$

and

$$K_{SPM} = -\hbar z |\beta_p|^2 k_{\omega_p} b_p^* b_p \qquad (37)$$

where $\beta(\omega_p) \equiv \beta_p$ and $\Delta k = 2k_{\omega_p} - k_{\omega_s} - k_{\omega_i}$ are the chromatic dispersions.

Similar to the previous classical treatment, one can introduce the co-rotating framework by replacing the field operators with

$$\hat{a}_{R\omega_s}(z) = \hat{\tilde{a}}_{R\omega_s}(z)\ e^{2i|\beta_p|^2 k_{\omega_s} z} \quad \text{and} \quad b_p(z) = \tilde{b}_p(z)\ e^{i|\beta_p|^2 k_{\omega_p} z} \qquad (38)$$

In this framework, one can derive the following differential equation

$$\frac{d\hat{\tilde{a}}_{R\omega_s}(z)}{dz} = \frac{i}{\hbar}\left[\frac{d}{dz}\hat{K}_1, \hat{\tilde{a}}_{R\omega_s}\right] = i\beta_p^2 \sqrt{k_{\omega_s} k_{\omega_i}} e^{i\Psi_2(\omega_s)z} \hat{\tilde{a}}_{R\omega_i}^\dagger \qquad (39)$$

and

$$\frac{\partial \tilde{b}_p(z)}{\partial z} = -\left\{ \frac{dK_{SPM}}{dz}, \tilde{b}_p \right\} = 0 \tag{40}$$

where $\Psi_2(\omega_s) = \Delta k + 2\left|\beta_p\right|^2 \left(k_{\omega_p} - k_{\omega_s} - k_{\omega_i}\right)$ is the total phase mismatch. These latter are formally identical to Eqs. (12) and (13), up to a frequency-dependent normalization of the wave-amplitudes, under the undepleted pump approximation. Reference [11] derives an exact solution for Eq. (39), being

$$\hat{\tilde{a}}^{out}_{R\omega_s}(z) = \left[\left(\cosh\left(g_2(\omega_s)z\right) - \frac{i\Psi_2(\omega_s)}{2g_2(\omega_s)} \sinh\left(g_2(\omega_s)z\right) \right) \hat{\tilde{a}}_{R\omega_s} \right.$$
$$\left. +i\frac{\beta_p^2\sqrt{k_{\omega_s}k_{\omega_i}}}{g_2(\omega_s)} \sinh\left(g_2(\omega_s)z\right) \hat{\tilde{a}}^\dagger_{R\omega_i} \right] e^{i\frac{\Psi_2(\omega_s)}{2}z} \tag{41}$$

where

$$g_2(\omega_s) = \sqrt{\left|\beta_p^2\right|^2 k_{\omega_s}k_{\omega_i} - \left(\frac{\Psi_2(\omega_s)}{2}\right)^2} \tag{42}$$

If a state moves inside a TWJPA of length $z = aN$, the power gain will be $G(\omega_s, aN) = \langle \hat{\tilde{a}}^{out}_{R\omega_s}(aN) | \hat{\tilde{a}}^{out\dagger}_{R\omega_s}(aN)\rangle / \langle \hat{\tilde{a}}_{R\omega_s} | \hat{\tilde{a}}^\dagger_{R\omega_s}\rangle$.

3.3. Quantum Hamiltonian model based on discrete-mode operators

An alternative approach for the derivation of the quantum dynamics of a TWJPA is the one proposed in [32]. In this model, the quantum Hamiltonian for a 4WM parametric amplifier is expressed as the integral, along an arbitrary quantization length l_q, of the linear energy density stored in each element of the circuit. The energy stored per unit length in a Josephson junction can be derived from Eq. (5) by dividing each term by the elementary cell length a and replacing $\Delta\Phi_n$ with its continuous counterpart. Instead, the energy stored per unit length in a capacitance C can be alternatively expressed in terms of flux difference $\Delta\Phi$ or stored charge Q as

$$U_C = \frac{1}{a}\int_{t_0}^{t} VI \; dt' = \frac{1}{a}\int_{t_0}^{t} \frac{d\Delta\Phi}{dt'} \; C\frac{d}{dt'}\left[\frac{d\Delta\Phi}{dt'}\right] \; dt' = \frac{1}{2}\frac{C}{a}\Delta\Phi(t)\frac{\partial^2\Delta\Phi}{\partial t^2} \tag{43}$$

$$= \frac{1}{a}\int_{t_0}^{t} \frac{Q}{C}\frac{dQ}{dt'} \; dt' = \frac{1}{2a}\frac{1}{C}Q^2(t) \tag{44}$$

under the assumption that $\Delta\Phi(t_0) = 0$ and $Q(t_0) = 0$.

Therefore, the quantum Hamiltonian of the system can be expressed, with an approximation to the first nonlinear order, as

$$\hat{H} = \int_{l_q} \left[U_J + U_{C_J} + U_{C_0} \right] dx \approx$$
$$\approx \int_{l_q} \left[\left(\frac{1}{2aL_{J_0}}\Delta\hat{\Phi} - \frac{1}{24aL_{J_0}}\left(\frac{2\pi}{\Phi_0}\right)^2 \Delta\hat{\Phi}^3 + \frac{1}{2}\frac{C_J}{a}\frac{\partial^2\Delta\hat{\Phi}}{\partial t^2} \right)\Delta\hat{\Phi} + \frac{1}{2a}\frac{1}{C_0}\hat{Q}^2_{C_0} \right] dx \tag{45}$$

where $\hat{Q}$ and $\hat{\Phi}$ are quantum operators. The former can be expressed, as suggested in [34] and adapted for discrete-mode operators in [36], as

$$\hat{Q}_{C_0} = \sum_n \frac{C_0}{a} \hat{V}_{C_0,n} = \sum_n \frac{C_0}{a} \sqrt{\frac{\hbar \omega_n a}{2 C_0 l_q}} \left(\hat{a}_n \, e^{i(k_n x - \omega_n t)} + \text{H.c.} \right) \qquad (46)$$

(here $k_{-n} = -k_n$ and $\omega_{-n} = \omega_n$).

Before defining the flux operator, it is necessary to define an effective inductance L_{eff} of the transmission line (modeled, as shown in **Figure 2**, as a parallel of the nonlinear Josephson inductance L_J and the capacitance C_J):

$$\frac{1}{j\omega_n L_{eff}} = \frac{1}{j\omega_n L_J} + j\omega_n C_J \quad \text{hence} \quad L_{eff} = \frac{L_J}{1 - \omega^2 L_J C_J} \equiv L_J \Lambda_n \qquad (47)$$

Using the telegrapher's equation [37], the discrete-mode current operator, under slowly varying amplitude approximation $(\partial \hat{a}_n / \partial x \approx 0)$, can be derived from the discrete-mode voltage operator as:

$$\hat{I}_{L_{eff}} = \sum_n \text{sgn}(n) \sqrt{\frac{\hbar \omega_n a}{2 L_{eff} l_q}} \left(\hat{a}_n \, e^{i(k_n x - \omega_n t)} + \text{H.c.} \right) \qquad (48)$$

Therefore, the flux operator can be expressed as

$$\Delta \hat{\Phi} = \frac{L_J}{a} \hat{I}_{L_{eff}} \quad \text{where} \quad L_J(\Delta \Phi) = \frac{\Delta \Phi}{I_J} = \frac{\frac{2\pi}{\Phi_0} \Delta \Phi}{I_c \frac{2\pi}{\Phi_0} \sin\left(\frac{2\pi}{\Phi_0} \Delta \Phi\right)} = L_{J_0} \frac{\frac{2\pi}{\Phi_0} \Delta \Phi}{\sin\left(\frac{2\pi}{\Phi_0} \Delta \Phi\right)} \qquad (49)$$

The recursive relation deriving from Eq. (49) can be solved iteratively. Exploiting a power series expansion of the sine function and considering just the first order of interaction, it results that

$$\Delta \hat{\Phi} = \sum_n \left[\left[1 + \frac{\Lambda_n}{12} \left(\frac{2\pi}{\Phi_0} \Delta \hat{\Phi}^{(0)} \right)^2 + O\left[\left(\frac{2\pi}{\Phi_0} \Delta \hat{\Phi}^{(0)} \right)^4 \right] \right] \Delta \hat{\Phi}_n^{(0)} \qquad (50)$$

where $\Delta \hat{\Phi}^{(0)}$ is the zero-order approximation of the flux quantum operator

$$\Delta \hat{\Phi}^{(0)} = \sum_n \Delta \hat{\Phi}_n^{(0)} = \sum_n \frac{k_n a}{\omega_n} \sqrt{\frac{\hbar \omega_n a}{2 C_0 l_q}} \left(\hat{a}_n \, e^{i(k_n x - \omega_n t)} + \text{H.c.} \right) \qquad (51)$$

Substituting Eqs. (46) and (50) in Eq. (45), and limiting the expression to the first nonlinear order, the Hamiltonian for a 4WM amplifiers turns up

$$\hat{H} = \sum_n \hbar \omega_n \left(\hat{a}_n^\dagger \hat{a}_n + \frac{1}{2} \right) + \sum_{n,m,l,k} \frac{-i\hbar^2 a}{96 L_{J_0} I_c^2 l_q^2 \Delta k_{nmlk}} e^{-i\Delta\omega_{nmlk}t} \left(e^{i\Delta k_{nmlk} l_q} - 1 \right)$$

$$\times \left\{ (1 - 4L_{J_0} \Lambda_n C_J \omega_k^2) \left(\hat{a} + \text{H.c.} \right)_{n \times m \times l \times k} + 4 L_{J_0} \Lambda_n C_J \left[2 \left(\omega \left(-i\hat{a} + \text{H.c.} \right) \right)_{n \times m} \right. \right.$$

$$\left. \left. \times \left(\hat{a} + \text{H.c.} \right)_{l \times k} + \left(\hat{a} + \text{H.c} \right)_{n \times m} \left(\omega \left(-i\hat{a} + \text{H.c.} \right) \right)_{l \times k} \right] \right\}$$

$$\qquad (52)$$

where $\hat{\bar{a}} \equiv \mathrm{sgn}\,(n)\sqrt{\Lambda_n \omega_n}\ \hat{a}_n$, $\Delta k_{nmlk} \equiv \pm k_n \pm k_m \pm k_l \pm k_k$, $\Delta \omega_{nmlk} \equiv \pm \omega_n \pm \omega_m \pm \omega_l \pm \omega_k$ (a $\pm$ sign refers to a corresponding annihilation (creation) operator) and the subscript $i \times j$ indicates a multiplication (i.e., $(\Lambda \omega)_{i \times j} = \Lambda_i \omega_i \Lambda_j \omega_j$).

Neglecting the constant zero-point energy and assuming a strong degenerate classical pump (as shown in [36])

$$\hat{a}_p \to -i\sqrt{\frac{\omega_p C_0 l_q}{2\hbar a}} A_p \tag{53}$$

it is possible to approximate the Hamiltonian in Eq. (52) to the second order in $\hat{a}_{s,i}^{(\dagger)}$ as

$$\hat{H}^{(CP)} \approx \sum_{n=s,i} \hbar \left(\omega_n + \xi_n |A_p|^2 \right) \hat{a}_n^\dagger \hat{a}_n - \hbar \left(\chi A_p^2 \hat{a}_s^\dagger \hat{a}_i^\dagger + \mathrm{H.c.} \right) \tag{54}$$

where

$$\xi_n = \frac{k_p^2 a^2 \Lambda_n \omega_n}{32 I_c^2 L_{J_0}^2} \left(4 - 3\delta_{pn} \right) \left(1 + \frac{2}{3}\left(\frac{\Lambda_p}{\Lambda_n} + \frac{\Lambda_n}{\Lambda_p} - 2 \right) \right) \tag{55}$$

represents the quantum self-phase modulation (when $n = p$) and the quantum cross-phase modulation (when $n = s, i$), whereas the coupling constant χ is defined as

$$\chi = \frac{k_p^2 a^2 \sqrt{\Lambda_s \omega_s \Lambda_i \omega_i}}{16 I_c^2 L_{J_0}^2} \left(1 + \frac{L_{J_0} C_J}{6} \left[\omega_p \omega_s \left(-2\Lambda_p + 5\Lambda_s - 3\Lambda_i \right) \right. \right.$$
$$\left. \left. + \omega_p \omega_i \left(-2\Lambda_p - 3\Lambda_s + 5\Lambda_i \right) + \omega_s \omega_i \left(4\Lambda_p - 2\Lambda_s - 2\Lambda_i \right) \right] \right) \tag{56}$$

Starting from the Hamiltonian $\hat{H}^{(CP)}$, it is possible to calculate the Heisenberg equation of motion for the classical pump amplitude and for the quantum operators $\hat{a}_s$ and $\hat{a}_i$, obtaining the coupled mode equations:

$$\frac{\partial A_p}{\partial t} = -i\left(\omega_p + 2\xi_p |A_p|^2 \right) A_p + 2i\chi^* A_p^* \hat{a}_s \hat{a}_i \tag{57}$$

$$\frac{\partial \hat{a}_{s(i)}}{\partial t} = -i\left(\omega_{s(i)} + \xi_{s(i)} |A_p|^2 \right) \hat{a}_{s(i)} + i\chi A_p^2 \hat{a}_{i(s)}^\dagger \tag{58}$$

In [32], the hypothesis under which the classical coupled mode equations Eqs. (12) and (13) can be obtained from Eqs. (57) and (58) is described in detail.

Moving to a co-rotating frame $\left(\hat{a}_{s(i)} \to \hat{a}_{s(i)} e^{i\xi_{s(i)}|A_{p_0}|^2 z} \right)$, the Hamiltonian (Eq. (54)) can be expressed as

$$\hat{H}_{rot}^{CP} = -\hbar \left(\chi |A_p|^2 \hat{a}_s^\dagger \hat{a}_i^\dagger e^{-i\Psi_3' t} + \mathrm{H.c.} \right) \tag{59}$$

where $\Psi_3' = \left(4\xi_p - \xi_s - \xi_i \right)|A_{p_0}|^2$. In this frame, introducing the undepleted pump assumption (Eq. (58)) turns into

$$\frac{\partial \hat{a}_{s(i)}}{\partial t} = i\chi |A_{p_0}|^2 \hat{a}_{i(s)}^\dagger e^{-i\Psi_3' t} \tag{60}$$

whose solutions are

$$\hat{a}_{s(i)}(t) = \left[\hat{a}_{s(i)_0} \left(\cosh\left(g_3' t\right) + \frac{i\Psi_3'}{2g_3'} \sinh\left(g_3' t\right) \right) + \frac{i\chi |A_{p_0}|^2}{g_3'} \hat{a}_{i(s)_0}^{\dagger} \sinh\left(g_3' t\right) \right] \quad (61)$$

where the exponential complex gain factor is defined as

$$g_3' = \sqrt{|\chi|^2 |A_{p_0}|^4 - \left(\frac{\Psi_3'}{2}\right)^2} \quad (62)$$

If a state spends a time t in the amplifier, the gain can be expressed as $G_s^Q(t) = \langle \hat{a}_s^{\dagger}(t)\hat{a}_s(t)\rangle / \langle \hat{a}_{s_0}^{\dagger}\hat{a}_{s_0}\rangle$.

To make the results of this last treatment, in which the operators are expressed as a function of the time, comparable with the previous ones, in which the operators are expressed as a function of the space coordinate, we need to take into account the phase velocity of the tones. It turns out that:

$$\Psi_3 = \Delta k + \left(4\xi_p \frac{\omega_p}{|k_p|} - \xi_s \frac{\omega_s}{|k_s|} - \xi_i \frac{\omega_i}{|k_i|} \right) \quad \text{and} \quad g_3 = \sqrt{|\chi|^2 |A_{p_0}|^4 \left(\frac{\omega_p}{|k_p|}\right)^2 - \left(\frac{\Psi_3}{2}\right)^2}$$

$$(63)$$

where $\Delta k = 2k_p - k_s - k_i$ is the chromatic dispersion.

3.4. Models comparison

The three exponential complex gain factors (g_i) and the three total phase mismatches (Ψ_i) derived in these models are analytically different but numerically similar, as shown in **Figure 3** (where the two insets report the differences between the quantum predictions and the classical ones).

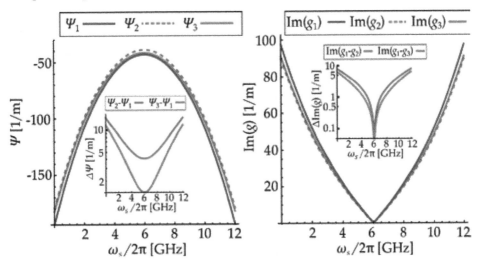

Figure 3.
Comparison of total phase mismatches (Ψ_i) and exponential complex gain factors (g_i) predicted by the three treatments presented in Section 3. For these calculations, typical constructive parameters have been assumed: $a = 50 \ \mu m$, $I_c = 5 \ \mu A$, $C_J = 300 \ fF$, and $C_0 = 35 fF$, in such a way that the characteristic impedance is $Z \approx 50 \ \Omega$. Furthermore $\omega_p/2\pi = 6$ GHz and $I_p = Ic/2$. In the insets, the differences between the quantum predictions and the classical ones are presented.

It is important to observe that in Eqs. (17), (41), and (61), the last term is always equal to zero in the case of a zero initial idler amplitude. In such a case, and under the hypothesis of a perfect phase matching ($\Psi_i = 0$), g_i is real and the amplification gain increases exponentially with the line length; whereas, in the case of a nonzero phase mismatch, g_i is imaginary and the gain increases quadratically [11].

4. Conclusions

In the present chapter, we have presented the state-of-the-art of the experimental evidences in the field of Josephson junctions-based traveling-wave metamaterials through a historical review in Section 2. Moreover, in Section 3, we have reported three different theoretical approaches for the prediction of a TJWPA dynamics, in the particular case of a 4WM process. Assuming similar simplifying hypothesis, like the use of a classical undepleted degenerate pump, the presence of slowly varying fields along the transmission line and approximating the non-linearities of the system up to the first order, a similar expression for the signal amplitude (or field annihilation/creation operators in the case of quantum theories) expressed in a co-rotating frame, is derived in the three treatments. Although the results of the quantum theories are similar to the classical ones, the description of the system dynamics with a quantum theory grants the possibility to evaluate photon-number distributions, squeezing effects and averages, standard deviations or higher-order moments of the measurements operators, taking into account the commutation relations between operators explicitly. For instance, detailed calcula-tions of the output state of a TWJPA in the case of a single-photon input state and in the case of a coherent input state are presented in [32].

Acknowledgements

The author would like to thank Luca Callegaro for the stimulating discussion. This work was partially funded by the Joint Research Project PARAWAVE of the European Metrology Programme for Innovation and Research (EMPIR). This project has received funding from the EMPIR program co-financed by the Participating States and from the European Unions' Horizon 2020 research and innovation program.

Author details

Luca Fasolo[1,2†], Angelo Greco[1,2†] and Emanuele Enrico[2*]

1 Politecnico di Torino, Torino, Italy

2 INRiM - Istituto Nazionale di Ricerca Metrologica, Torino, Italy

*Address all correspondence to: e.enrico@inrim.it

† These authors contributed equally.

References

[1] Devoret MH, Schoelkopf RJ. Superconducting circuits for quantum information: An outlook. Science. 2013;**339**:1169. DOI: 10.1126/science.1231930

[2] Josephson BD. Possible new effects in superconductive tunneling. Physics Letters. 1962;**1**(7):251-253. DOI: 10.1016/0031-9163(62)91369-0

[3] Niemczyk T et al. Circuit quantum electrodynamics in the ultrastrong-coupling regime. Nature Physics. 2010;**6**:772776. DOI: 10.1038/NPHYS1730

[4] Lamb WE, Retherford RC. Fine structure of the hydrogen atom by a microwave method. Physics Review. 1947;**72**:241. DOI: 10.1103/PhysRev.72.241

[5] Fragner A, Goppl M, Fink JM, Baur M, Bianchetti R, Leek PJ, et al. Resolving vacuum fluctuations in an electrical circuit by measuring the Lamb shift. Science. 2008;**322**:1357. DOI: 10.1126/science.1164482

[6] Moore GT. Quantum theory of the electromagnetic field in a variable-length one-dimensional cavity. Journal of Mathematical Physics. 1970;**11**:2679. DOI: 10.1063/1.1665432

[7] Fulling SA, Davies PCW. Radiation from a moving mirror in two dimensional space-time: Conformal anomaly. Proceedings of the Royal Society A. 1976;**348**:393. DOI: 10.1098/rspa.1976.0045

[8] Wilson C et al. Observation of the dynamical Casimir effect in a superconducting circuit. Nature. 2011;**479**:376379. DOI: 10.1038/nature10561

[9] Ketchen MB et al. Subm, planarized, NbAlOxNb Josephson process for 125 mm wafers developed in partnership with Si technology. Applied Physics Letters. 1991;**59**:2609. DOI: 10.1063/1.106405

[10] White TC, al e. Traveling wave parametric amplifier with Josephson junctions using minimal resonator phase matching. Applied Physics Letters. 2015;**106**:242601. DOI: 10.1063/1.4922348

[11] O'Brien K, Macklin C, Siddiqi I, Zhang X. Resonant phase matching of Josephson junction traveling wave parametric amplifiers. Physical Review Letters. 2014;**113**:157001. DOI: 10.1103/PhysRevLett.113.157001

[12] Cullen AL. Theory of the travelling-wave parametric amplifier. Proceedings of the IEE - Part B: Electronic and Communication Engineering. 1959;**32**(107):101-107. DOI: 10.1049/pi-b-2.1960.0085

[13] Sweeny M, Mahler R. A travelling-wave parametric amplifier utilizing Josephson junctions. IEEE Transactions on Magnetics. 1985;**21**(2):654-655. DOI: 10.1109/TMAG.1985.1063777

[14] Macklin C, OBrien K, Hover D, Schwartz ME, Bolkhovsky V, Zhang X, et al. A nearquantum-limited Josephson traveling-wave parametric amplifier. Science. 2015;**350**(6258):307-310. DOI: 10.1126/science.aaa8525

[15] Walls DF, Milburn GJ. Quantum Optics. 2nd ed. Berlin: Springer; 2008

[16] Einstein A, Podolsky B, Rosen N. Can quantum-mechanical description of physical reality be considered complete?Physics Review. 1935;**47**:777. DOI: 10.1103/PhysRev.47.777

[17] Reid MD, Drummond PD. Quantum correlations of phase in nondegenerate parametric oscillation. Physical Review Letters. 1988;**60**:2731. DOI: 10.1103/PhysRevLett.60.2731

[18] Butikov EI. Parametric resonance in a linear oscillator at square-wave modulation. European Journal of Physics. 2005;**26**:157174. DOI: 10.1088/ 0143-0807/26/1/016

[19] Mavaddat R, Hyde FJ. Investigation of an experimental travelling-wave parametric amplifier. Proceedings of the IEE - Part B: Electronic and Communication Engineering. 1962; **109**(47):405. DOI: 10.1049/pi-b- 2.1962. 0225

[20] Yurke B, Roukes ML, Movshovich R, Pargellis AN. A low noise series array Josephson junction parametric amplifier. Applied Physics Letters. 1996;**69**:3078. DOI: 10.1063/ 1.116845

[21] van der Zant HSJ, Berman D, Orlando TP. Fiske modes in one-dimensional parallel Josephson-junction arrays. Physical Review B. 1994;**49**:18. DOI: 10.1103/PhysRevB.49.12945

[22] Caputo P, Darula M, Ustinov AV, Kohlstedt H. Fluxon dynamics in discrete Josephson transmission lines with stacked junctions. Journal of Applied Physics. 1997;**81**:309. DOI: 10.1063/1.364110

[23] Mohebbi HR, Hamed Majedi A. Analysis of series-connected discrete Josephson transmission line. IEEE Transactions on Microwave Theory and Techniques. 2009;**57**:8. DOI: 10.1109/TMTT.2009.2025413

[24] Yaakobi O, Friedland L, Macklin C, Siddiqi I. Parametric amplification in Josephson junction embedded transmission lines. Physical Review B. 2013;**87**:144301. DOI: 10.1103/ PhysRevB.87.144301

[25] Tan BK, Yassin G. Design of a uniplanar resonance phase-matched Josephson travelling-wave parametric amplifier. In: 10th UK-Europe-China Workshop on Millimetre Waves and Terahertz Technologies (UCMMT); September 11–13, 2017; Liverpool. New York: IEEE; 2017. pp. 1-4

[26] Bell MT, Samolov A. Traveling-wave parametric amplifier based on a chain of coupled asymmetric SQUIDs. Physical Review Applied. 2015;**4**: 024014. DOI: 10.1103/PhysRevApplied.4.024014

[27] Zhang W, Huang W, Gershenson ME, Bell MT. Josephson metamaterial with a widely tunable positive or negative Kerr constant. Physical Review Applied. 2017;**8**: 051001. DOI: 10.1103/ PhysRevApplied.8.051001

[28] Zorin AB. Josephson traveling-wave parametric amplifier with three-wave mixing. Physical Review Applied. 2016; **6**:034006. DOI: 10.1103/PhysRev Applied.6.034006

[29] Zorin AB, Khabipov M, Dietel J, Dolata R. Traveling-wave parametric amplifier based on three-wave mixing in a Josephson Metamaterial. In: 2017 16th International Superconductive Electronics Conference (ISEC); June 12–16, 2017; Naples. New York: IEEE; 2017. pp. 1-3

[30] Miano A, Mukhanov OA. Symmetric traveling wave parametric amplifier. IEEE Transactions on Applied Superconductivity. 2019;**29**:5. DOI: 10.1109/TASC.2019.2904699

[31] Grimsmo AL, Blais A. Squeezing and quantum state engineering with Josephson travelling wave amplifiers. npj Quantum Information. 2017;**3**:20. DOI: 10.1038/s41534-017-0020-8

[32] van der Reep THA. Mesoscopic Hamiltonian for Josephson traveling-wave parametric amplifiers. Physical Review A. 2019;**99**:063838. DOI: 10.1103/PhysRevA.99.063838

[33] Roy A, Devoret M. Quantum-limited parametric amplification with Josephson circuits in the regime of pump depletion. Physical Review B. 2018;**98**:045405. DOI: 10.1103/PhysRevB.98.045405

[34] Vool U, Devoret M. Introduction to quantum electromagnetic circuits. International Journal of Circuit Theory and Applications. 2017;**45**:897. DOI: 10.1002/cta.2359

[35] Santos DJ, Loudon R. Electromagnetic-field quantization in inhomogeneous and dispersive one-dimensional systems. Physical Review A. 1995;**52**:1538. DOI: 10.1103/PhysRevA.52.1538

[36] Loudon R. The Quantum Theory of Light. 3rd ed. Oxford, UK: Oxford University Press; 2000

[37] Pozar DM. Microwave Engineering. 4th ed. Hoboken, NJ: Wiley; 2012

2

G-Jitter Effects on Chaotic Convection in a Rotating Fluid Layer

Palle Kiran

Abstract

The effect of gravity modulation and rotation on chaotic convection is investigated. A system of differential equation like Lorenz model has been obtained using the Galerkin-truncated Fourier series approximation. The nonlinear nature of the problem, i.e., chaotic convection, is investigated in a rotating fluid layer in the presence of g-jitter. The NDSolve Mathematica 2017 is employed to obtain the numerical solutions of Lorenz system of equations. It is found that there is a proportional relation between Taylor number and the scaled Rayleigh number R in the presence of modulation. This means that chaotic convection can be delayed (for increasing value of R) or advanced with suitable adjustments of Taylor number and amplitude and frequency of gravity modulation. Further, heat transfer results are obtained in terms of finite amplitude. Finally, we conclude that the transition from steady convection to chaos depends on the values of Taylor number and g-jitter parameter.

Keywords: g-jitter effect, nonlinear theory, rotation, chaos, truncated Fourier series

1. Introduction

The study of chaotic convection is of great interest due to its applications in thermal and mechanical engineering and in many other industry applications. It was introduced by Lorenz [1] to illustrate the study of atmospheric three-space model arising from Rayleigh-Benard convection. Some of the applications are production of crystals, oil reservoir modeling, and catalytic packed bed filtration. He developed a simplified mathematical model for atmospheric convection given below:

$$x' = Pr(y - x), \tag{1}$$

$$y' = x(R - z) - y, \tag{2}$$

$$z' = xy - \beta z. \tag{3}$$

This model is a system of three ordinary differential equations known as the Lorenz equations. These equations are related to the properties of a two-dimensional Rayleigh-Benard convection. In particular, the system describes the rate of change of three quantities convection, temperature variation vertically with respect to time. These equations are related to the properties of two-dimensional flow model warmed uniformly from below and cooled from above. In particular, the system describes the rate of change of three quantities of time, x is proportional to the rate of convection, y is the horizontal temperature variation, and z is the vertical temperature variation. The constants Pr, R and β are the system parameters proportional to the Prandtl number, Rayleigh number, and certain physical dimensions of the media. If $R < 1$ then there is only one equilibrium point at the origin which is represented as no convection point. Further, all orbits converge to the origin, which is a global attractor. When $R = 1$, then a pitchfork bifurcation occurs, and for R_1, two additional critical points arise and are known as convection points, and there the system loses its stability. In addition to this model, I would like to add the concept of modulation either to suppress or to enhance nonlinearity. The literature shows that there are different types available; some of them are temperature modulation (Venezian [2]), gravity (Gresho and Sani [3] and Bhadauria and Kiran [4, 5]), rotation (Donnelly [6], Kiran and Bhadauria [7]), and magnetic field modulation (Bhadauria and Kiran [8, 9]). Their studies are mostly on thermal convection either considering fluid or porous medium. Their ultimate idea behind the research is to find external regulation to the system to control instability and measure the heat mass transfer in the system. But what happens when we consider the external configuration to system Eq. (1). The external configurations are like thermal, gravity, rotation, and magnetic field modulation. In this direction, no data are reported so far. With this, I would like to extend the work of Lorenz along with modulation.

The studies on chaos with respect to the different types of parameters like Rayleigh number and Prandtl number are mostly investigated by the following studies. The transition from steady convection to chaos occurs by a subcritical Hopf bifurcation producing a solitary cycle which may be associated with a homoclinic explosion for low Prandtl number is investigated by Vadasz and Olek [10]. The work of Vadasz [11] suggests an explanation for the appearance of this solitary limit cycle via local analytical results. The effect of magnetic field on chaotic convection in fluid layer is investigated by Mahmud and Hasim [12]. They found that transition from chaotic convection to steady convection occurs by a subcritical Hopf bifurcation producing a homoclinic explosion which may limit the cycle as Hartman number increases. For the moderate values of Prandtl number, the route to chaos occurs by a period of doubling sequence of bifurcations given by Vadasz and Olek [13]. Feki [14] proposed a new simple adaptive controller to control chaotic systems. The constructed linear structure of controller may be used for chaos control as well as for chaotic system synchronization. Yau and Chen [15] found that the Lorenz model could be stabilized, even in the existence of system external distraction. For non-Newtonian fluid case, Sheu et al. [16] have shown that stress relaxation tends to accelerate onset chaos. A weak nonlinear solution to the problem is assumed by Vadasz [17], and it can produce an accurate analytical expression for the transition point as long as the condition of validity and consequent accuracy of the latter solution is fulfilled. Narayana et al. [18] investigated heat mass transfer using truncated Fourier series method. They have also discussed chaotic convection under the effect of binary viscoelastic fluids. The studies related to gravity modulation are given by Kiran et al. [19–25]. These studies show that the gravity modulation can be used to control heat and mass transfer in the system in terms of frequency and amplitude of modulation.

The above paragraph demonstrated the earlier work on chaotic convection with different configurations and models to control chaos. Recently Vadasz et al. [26] and Kiran et al. [27] have investigated the effect of vertical vibrations and temperature modulation on chaos in a porous media. Their results show that periodic solutions and chaotic solutions alternate as the value of the scaled Rayleigh number changes in the presence of forced vibrations. The root to chaos is also affected by three types of thermal modulations.

The effect of rotation on chaos is investigated by Gupta et al. [28] without any modulation. They found that rotation has delay in chaos and controls nonlinearity. It is also concluded that there are suitable ranges over Ta and R to reduce chaos in the system. Based on the above studies in this chapter, I would like to investigate the study of chaotic convection in the presence of rotation and gravity modulation.

2. Mathematical model

An infinitely extended horizontal rotating fluid layer about its vertical z-axis is considered. The layer is gravity modulated and the lower plate held at temperature T_0 while the upper plate at $T_0 + \Delta T$. Here ΔT is the temperature difference in the medium. The mathematical equation of the flow model is given by

$$\nabla . q = 0, \tag{4}$$

$$\frac{\partial \overline{q}}{\partial t} + 2\Omega * \overline{q} = -\frac{1}{\rho_0}\nabla p + \frac{\rho}{\rho_0}\overline{g} + \nu \Delta^2 \overline{q}, \tag{5}$$

$$\frac{\partial T}{\partial t} + (\overline{q}.\nabla)T = k_T \nabla^2 T, \tag{6}$$

$$\rho = \rho_0 [1 - \alpha_T (T - T_0)]. \tag{7}$$

The thermal boundary conditions are given by

$$T = T_0 + \Delta T \quad at \quad z = 0 \quad and \quad T = T_0 \quad at \quad z = d, \tag{8}$$

where $\overline{q} - >$ is the velocity of the fluid, $\Omega - >$ is the vorticity vector, $p - >$ is the fluid pressure, $\rho - >$ is the density, $\nu - >$ is the kinematic viscosity, $K_T - >$ is the thermal diffusivity ratio, and $\alpha_t - >$ is the thermal expansion coefficient. We consider in our problem the externally imposed gravitational field (given by Gresho and Sani [3]):

$$\vec{g} = g_0 [1 + \delta_g \sin(\omega_g t)]\hat{k}, \tag{9}$$

where δ_g, ω_g are the amplitude and frequency of gravity modulation.

2.1 Basic state

The basic state of the fluid is quiescent and is given by

$$q_b = (0,0,0), p = p_b(z), T = T_b(z). \tag{10}$$

Using the basic state Eq. (10) in the Eqs. (4)–(6), we get the following relations

$$\frac{\partial \overline{q}_b}{\partial t} + 2\Omega * \overline{q}_b = -\frac{1}{\rho_0}\nabla p_b + \frac{\rho_b}{\rho_0}\overline{g} + \nu \Delta^2 \overline{q}_b, \tag{11}$$

$$o = -\frac{1}{\rho_0}\nabla p_b + \frac{\rho_b}{\rho_0}\overline{g}, \tag{12}$$

$$\nabla p_b = \rho_b \overline{g}, \tag{13}$$

$$\frac{\partial p_b}{\partial z} = \rho_b \overline{g}, \tag{14}$$

and from Eq. (6)

$$\frac{\partial T_b}{\partial t} + (\bar{q}_b.\nabla)T = k_T\nabla^2 T_b, \tag{15}$$

$$k_T\nabla^2 T_b = 0, \tag{16}$$

$$T_b = T_0 + \Delta T\left(1 - \frac{z}{d}\right). \tag{17}$$

2.2 Perturbed state

On the basic state, we superpose perturbations in the form

$$q = q_b + q', \rho = \rho_b(z) + \rho', p = p_b(z) + p', T = T_b(z) + T' \tag{18}$$

where the primes denote perturbed quantities. Now substituting Eq. (18) into Eqs. (4)–(7) and using the basic state solutions, we obtain the equations governing the perturbations in the form

$$\nabla.\bar{q}' = , 0 \tag{19}$$

$$\frac{\partial(T_b + T')}{\partial t} + \left((q_b + q').\Delta\right)(T_b + T') = K_T\nabla^2(T_b + T'), \tag{20}$$

$$\frac{\partial T'}{\partial t} + (q'.\nabla)(T_b + T') = K_T\nabla^2(T'), \tag{21}$$

$$\frac{\partial T'}{\partial t} + \left(u'\frac{\partial}{\partial x} + w'\frac{\partial}{\partial z}\right)(T_b + T') = K_T\nabla^2(T'), \tag{22}$$

simplifying the above equation, then we get

$$\frac{\partial T'}{\partial t} - \frac{\partial \psi}{\partial x}\frac{\partial T_b}{\partial z} + \frac{\partial(\psi, T')}{\partial(x,z)} = K_T\nabla^2(T'). \tag{23}$$

Similarly we can derive the same for momentum equation of the following form

$$\frac{\partial \bar{q}'}{\partial t} + 2\Omega * \bar{q}' = -\frac{1}{\rho_0}\nabla p' + \frac{\rho'}{\rho_0}\bar{g} + \nu\Delta^2\bar{q}'. \tag{24}$$

We consider only two-dimensional disturbances and define the stream functions ψ and $\bar{q}$ by

$$(u', w') = \left(-\frac{\partial \psi}{\partial z}, \frac{\partial \psi}{\partial x}\right), \bar{g} = (0, 0, -g), \tag{25}$$

which satisfy the continuity Eq. (19). While introducing the stream function ψ and non-dimensionalizing with the following nondimensional parameters $(x', y', z') = d(x^*, y^*, z^*)$, $t' = \frac{d^2}{K_T}t^*$, $T' = (\Delta T)T^*$, and $p' = \frac{\mu K_T}{d^2}p^*$, then the resulting Eq. (19) becomes

$$\frac{\partial T'}{\partial t} - \frac{\partial \psi}{\partial x}\frac{\partial T_b}{\partial z} + \frac{\partial(\psi, T')}{\partial(x,z)} = K_T\nabla^2(T'),$$

after simplifying the above equation, we get

$$\left(\frac{\partial}{\partial t} - \nabla^2\right) T = \frac{\partial \psi}{\partial x} - \frac{\partial(\psi, T)}{\partial(x, z)}. \tag{26}$$

Similarly while eliminating the pressure term and using the dimensionless quantities, from the momentum equation (24), we get the following:

$$\left[\left(\frac{1}{Pr}\frac{\partial}{\partial t} - \nabla^2\right)^2 \nabla^2 + T_a \frac{\partial^2}{\partial z^2}\right]\frac{\partial \psi}{\partial x} = Ra\left(1 + \delta_g \sin\left(\omega_g t\right)\right)\frac{\partial^2}{\partial x^2}\left(\frac{1}{Pr}\frac{\partial}{\partial t} - \nabla^2\right)T, \tag{27}$$

where Pr = $\frac{\nu}{K_T}$ is the Prandtl number, $T_a = \frac{4d^4\Omega^2}{\nu^2}$ is the Taylor number, and $Ra = \frac{\alpha(\Delta T)d^3 g_0}{\nu K_T}$ is the Rayleigh number. The assumed boundaries are stress free and iso-thermal; therefore, the boundary conditions are given by

$$w = \frac{\partial^2 w}{\partial z^2} = T = 0 \quad at \quad z = 0 \quad and \quad z = 1. \tag{28}$$

The set of partial differential Eqs. (26) and (27) forms a nonlinear coupled system of equations involving stream function and temperature as a function of two variables in x and z. We solve these equations by using the Galerkin method and using Fourier series representation.

3. Truncated Galerkin expansion

To obtain the solution of nonlinear coupled system of partial differential equations (26) and (27), we represent the stream function and temperature in the form

$$\psi = A_1 \sin(ax)\sin(\pi z), \tag{29}$$

$$T = B_1 \cos(ax)\sin(\pi z) + B_2 \sin(2\pi z) \tag{30}$$

The above are the Galerkin expansion of stream function and temperature. Now substituting these equations in Eqs. (26) and (27) and applying the orthogonal conditions to Eqs. (30) and (31) and finally integrating over the domain [0,1] × [0,1] yield a set of equations:

$$\frac{\partial B_1}{\partial t}\cos ax \sin \pi z + \frac{\partial B_2}{\partial t}\sin 2\pi z + k^2 B_1 \cos ax \sin \pi z + 4B_2\pi^2 \sin 2\pi z \tag{31}$$

$$= A_1 a \cos ax \sin \pi z - A_1 B_1 a\pi \cos \pi z \sin \pi z \tag{32}$$

$$-2A_1 B_2 a\pi \cos 2\pi z \cos ax \sin \pi z. \tag{33}$$

Now multiply with $\cos ax \sin \pi z$ on both sides, and apply integration from 0 to 1 with respect to x and 0 to $\frac{2\pi}{a}$:

$$\frac{\partial B_1}{\partial t}\int_0^1\int_0^{\frac{2\pi}{a}}\cos^2 ax \sin^2 \pi z\, dx dz + \frac{\partial B_2}{\partial t}\int_0^1\int_0^{\frac{2\pi}{a}}\cos ax \sin \pi z \sin 2\pi z\, dx dz \tag{34}$$

$$+k^2 B_1\int_0^1\int_0^{\frac{2\pi}{a}}\cos^2 ax \sin^2 \pi z\, dx dz \tag{35}$$

$$+4B_2\pi^2\int_0^1\int_0^{\frac{2\pi}{a}} sin\,2\pi z\,\cos ax\,\sin \pi z dx dz \tag{36}$$

$$= A_1 a\int_0^1\int_0^{\frac{2\pi}{a}} \cos^2 ax\,\sin^2 \pi z dx dz \tag{37}$$

$$-A_1 B_1 a\pi\int_0^1\int_0^{\frac{2\pi}{a}} \cos ax\,\cos \pi z\,\sin^2 \pi z dx dz \tag{38}$$

$$-2A_1 B_2 a\pi\int_0^1\int_0^{\frac{2\pi}{a}} \cos 2\pi z\,\cos^2 ax\,\sin^2 \pi z dx dz. \tag{39}$$

$$\frac{\partial B_1}{\partial t}\frac{\pi}{2a}+k^2 B_1\frac{\pi}{2a}=A_1 a\frac{\pi}{2a}-2A_1 B_2 a\pi\left(-\frac{\pi}{2a}\right), \tag{40}$$

$$\frac{\partial B_1}{\partial t}=A_1 a+A_1 B_2 a\pi-k^2 B_1. \tag{41}$$

Now we consider $\tau=k^2 t\Rightarrow t=\frac{\tau}{k^2}$.

$$\frac{\partial B_1}{\partial \tau}=\frac{A_1 a}{k^2}+\frac{a\pi}{k^2}A_1 B_2-B_1. \tag{42}$$

Now let us consider Eq. (30) and multiply with $sin\,2\pi z$ on both sides of the equation and apply integration from 0 to 1 with respect to x and 0 to $\frac{2\pi}{a}$:

$$\frac{\partial B_1}{\partial t}\int_0^1\int_0^{\frac{2\pi}{a}} \cos ax\,\sin \pi z sin2\pi z dx dz+\frac{\partial B_2}{\partial t}\int_0^1\int_0^{\frac{2\pi}{a}} \sin^2 2\pi z dx dz \tag{43}$$

$$+k^2 B_1\int_0^1\int_0^{\frac{2\pi}{a}} \cos ax\,\sin \pi z\,\sin 2\pi z dx dz \tag{44}$$

$$+4B_2\pi^2\int_0^1\int_0^{\frac{2\pi}{a}} \sin^2 2\pi z dx dz \tag{45}$$

$$= A_1 a\int_0^1\int_0^{\frac{2\pi}{a}} \cos ax\,\sin \pi z\,\sin 2\pi z dx dz \tag{46}$$

$$-\int_0^1\int_0^{\frac{2\pi}{a}}-\int_0^1\int_0^{\frac{2\pi}{a}} A_1 B_1 a\pi\,\cos \pi z\,\sin \pi z\,\sin 2\pi z dx dz \tag{47}$$

$$-\int_0^1\int_0^{\frac{2\pi}{a}} 2A_1 B_2 a\pi cos2\pi z\,\cos ax\,\sin \pi z sin\,2\pi z dx dz, \tag{48}$$

then by simplifying the above equation, we get

$$\frac{\partial B_2}{\partial \tau}=-\frac{4\pi^2}{k^2}B_2-\frac{a\pi}{2k^2}A_1 B_1. \tag{49}$$

Similarly from Eq. (50)

$$\frac{\partial^2 A_1}{\partial \tau^2} = -2Pr\frac{\partial A_1}{\partial \tau} + \frac{a}{K^6}\left(a^2 Ra\left(1 + \delta_g \sin\left(\omega_g t\right)\right) - \pi^2 T_a Pr - k^6 Pr\right)A_1 + \frac{\pi a^2 Pr Ra}{k^6}A_1 B_2$$
$$+ \frac{aRaPr(Pr-1)}{k^4}B_1,$$

$$(50)$$

where $k^2 = \pi^2 + a^2$ is the total wavenumber and $\tau = k^2 t$ is the rescaled time. Introducing the following dimensionless quantities

$$R = \frac{a^2 Ra}{K^6}, T = \frac{\pi^2 T_a}{k^6} \ and \ \gamma = -\frac{4\pi^2}{k^2}, \sigma = Pr, \qquad (51)$$

and rescale the amplitudes in the form of

$$X = \frac{\pi a}{k^2\sqrt{2}}A_1, Y = \frac{\pi R}{\sqrt{2}}B_1 \quad and \quad Z = -\pi R B_2. \qquad (52)$$

To provide the following set of equations, we consider the following equations $\gamma = -\frac{4\pi^2}{k^2}, \frac{1}{k^2} = -\frac{\gamma}{4\pi^2}$

$$\frac{\partial B_1}{\partial \tau} == \frac{\gamma a}{4\pi^2}A_1 - \frac{\gamma a\pi}{4\pi^2}A_1 B_2 - B_1, \qquad (53)$$

$$\frac{\partial}{\partial \tau}\left(\frac{Y\sqrt{2}}{\pi R}\right) = \frac{\gamma aR}{4\pi^2}\left(\frac{Xk^2\sqrt{2}}{\pi aR}\right) - \frac{\gamma a}{4\pi}\left(\frac{Xk^2\sqrt{2}}{\pi a}\right)\left(-\frac{z}{\pi R}\right) - \frac{Y\sqrt{2}}{\pi R}, \qquad (54)$$

and then simplifying the above equation, we get

$$Y' = RX - XZ - Y, \qquad (55)$$

now from the Eq. (50)

$$\frac{\partial B_2}{\partial \tau} = \gamma B_2 - \frac{1}{2}\left(-\frac{\gamma}{4\pi^2}\right)\pi a A_1 B_1, \qquad (56)$$

$$\frac{\partial}{\partial \tau}\left(\frac{Z}{\pi R}\right) = \gamma\left(\frac{z}{\pi R}\right) - \frac{1}{2}\left(-\frac{\gamma}{4\pi^2}\right)\pi a\left(\frac{Xk^2\sqrt{2}}{\pi R}\right)\left(\frac{Y\sqrt{2}}{\pi R}\right), \qquad (57)$$

$$Z' = \gamma Z + XY. \qquad (58)$$

Similarly from Eq. (28),

$$X' = W, \qquad (59)$$

$$W' = -2\sigma w + \sigma\left(R\left(1 + \delta_g \sin\left(\omega_g t\right)\right) - \sigma(T+1)\right)X - \sigma XZ + \sigma(\sigma-1)Y, \qquad (60)$$

where the symbol ($/$) denotes the time derivative $\frac{d()}{d\tau}$. Eqs. (56), (59), and (61) are like the Lorenz equations (Lorenz (13), sparrow (14)), although with different coefficients. The final nonlinear differential equations are given by

$$X' = W, \tag{61}$$

$$Y' = RX - XZ - Y, \tag{62}$$

$$Z' = \gamma Z + XY, \tag{63}$$

$$W' = -2\sigma W + \sigma\left(R\left(1 + \delta_g \sin\left(\omega_g \tau\right)\right) - \sigma(T+1)\right)X - \sigma XZ + \sigma(\sigma - 1)Y. \tag{64}$$

4. Stability analyses

To understand the stability of the system, we determine the fixed points of the system and will try to find the nature of these fixed points through eigen equation. The nonlinear dynamics of Lorenz-like system (62)–(65) has been analyzed and solved for $\sigma = 10$, $\gamma = -\frac{8}{3}$ corresponding to convection. The basic properties of the system to obtain the eigen function are described next.

4.1 Dissipation

The system of Eqs. (62)–(65) is dissipative since

$$\nabla V = \frac{\partial X'}{\partial X} + \frac{\partial Y'}{\partial Y} + \frac{\partial Z'}{\partial Z} + \frac{\partial W'}{\partial W} = -(2\sigma + 1 - \gamma) < 0. \tag{65}$$

If the set of initial solutions is the region of $V(0)$, then after some time t, the endpoints of the trajectories will decrease to a volume:

$$V(t) = V(0)\,exp\left[-(2\sigma + 1 - \gamma)t\right]. \tag{66}$$

The above expression shows that the volume decreases exponentially with time.

4.2 Equilibrium points

System (62)–(65) has the general form, and the equilibrium (fixed or stationary) points are given by:

$$X' = W, \tag{67}$$

$$W = 0. \tag{68}$$

From Eq. (83) we got

$$X = \frac{Y}{R - Z}, \tag{69}$$

and similarly we also got the following from Eq. (64):

$$Z = \frac{-Y^2}{\gamma(R - Z)}, \tag{70}$$

and similarly we also got the following from Eq. (65) for the momentum case:

$$R = T + 1, \tag{71}$$

then we get a relation

$$X_{2,3} = \pm \frac{\sqrt{(T+1-R)\gamma}}{\sqrt{T+1}} \qquad (72)$$

the remaining $Y_{2,3}, Z_{2,3}$ will be accessed. The fixed points of rescaled system for modulated case are $(X_1, Y_1, Z_1) = (0, 0, 0)$ corresponding to the motionless solution and $(X_{2,3}, Y_{2,3}, Z_{2,3}) = \left[\pm\sqrt{\frac{z}{c}}, \pm c\sqrt{\frac{z}{c}}, \frac{(RI_1-c)}{(R-1)^2}\right]$ corresponding to the convection solution. The critical value of R, where the motionless solution loses their stability and the convection solution takes over, is obtained as $R_{cr} = \frac{c}{I_1}$, which corresponds to $Ra = 4\pi^2 \frac{c}{I_1}$ where $c = \left(1 + C\frac{\pi^2}{\gamma}\right)$ and $I_1 = \int_0^1 \sin^2(\pi z) f_2 dz$. This pair of equilibrium points is stable only if $R < \sqrt{\frac{z}{c}}$; beyond this condition the other periodic, quasi-periodic, or chaotic solutions take over at $R > \sqrt{\frac{z}{c}}$. The corresponding stability of the fixed points associated with the motionless solution $(X_1, Y_1, Z_1) = (0, 0, 0)$ is controlled by the zeros of the following characteristic polynomial:

5. Stability of equilibrium points

The Jacobian matrix of Eqs. (62)–(65) is as follows:

$$J = DF_{(X,Y,Z,W)} = \begin{bmatrix} 0 & 0 & 0 & 0 \\ R-Z & -1 & -X & 0 \\ Y & X & \gamma & 0 \\ \sigma[R - \sigma(T+1) - Z] & \sigma(\sigma-1) & -\sigma X & -2\sigma \end{bmatrix}.$$

The characteristic values of the above Jacobian matrix, obtained by solving the zeros of the characteristic polynomial, provide the stability conditions. If all the eigenvalues are negative, then the fixed point is stable (or in the case of complex eigenvalues, they have negative real parts) and unstable, when at least one eigenvalue is positive (or in the case of complex eigenvalues, it has positive real part):

$$DF_{(0,0,0,0)} = \begin{bmatrix} 0 & 0 & 0 & 0 \\ R & -1 & 0 & 0 \\ 0 & 0 & \gamma & 0 \\ \sigma[R - \sigma(T+1)] & \sigma(\sigma-1) & 0 & -2\sigma \end{bmatrix}.$$

The characteristic equation for the above system at origin is given by $|A - \lambda I| = 0$ which implies the following

$$\gamma = \lambda, \lambda^3 + (2\sigma+1)\lambda^2 + \left[(2-R)\sigma + \sigma^2(T+1)\right]\lambda + \sigma^2(T-R+1) = 0.$$

The first eigenvalue γ is always negative as $\gamma = \frac{-8}{3}$, but the other three eigenvalues are given by equation

$$\lambda^3 + (2\sigma + 1)\lambda^2 + \left[(2 - R)\sigma + \sigma^2(T + 1)\right]\lambda + \sigma^2(T - R + 1) = 0.$$

The stability of the fixed points corresponding to the convection solution $(X_{2,3}, Y_{2,3}, Z_{2,3})$ is controlled by the following equation for the eigenvalues $\lambda_i, = 1, 2, 3, 4$:

$$\lambda^4 + \lambda^3(2\sigma + 1 - \gamma) + \lambda^2\left(2\sigma - \gamma - 2\gamma\sigma\gamma - \sigma T + \sigma^2 T - \sigma + \sigma^2 + X^2\right) + \lambda\left(X^2\sigma(T + 1)\right) \tag{73}$$

$$-\sigma\gamma + T\sigma\gamma - \sigma^2\gamma T - \sigma^2\gamma) + 2X^2\sigma^2(T + 1) = 0, \tag{74}$$

$$\lambda^4 + \lambda^3(2\sigma + 1 - \gamma) + \lambda^2 + \left[\frac{-\gamma R}{T + 1} + 2\sigma(1 - \gamma) + \sigma(\sigma - 1)(T + 1)\right]\lambda^2 \tag{75}$$

$$+\left[\frac{-2\sigma\gamma R}{T + 1} + \sigma\gamma(2 - \sigma)(T + 1) - R\right]\lambda + 2\sigma^2 Y(T + 1 - R) = 0, \tag{76}$$

$$\frac{\sigma\gamma^2(T + 3)(1 - \gamma - \sigma - \sigma T)}{(T + 1)^2}R^2 - \sigma\gamma\left[(2\sigma + 1 - \gamma)\{\gamma(2 - \sigma) + \frac{2\sigma(1 - \gamma)(T + 3)}{T + 1}\right. \tag{77}$$

$$+\sigma(T + 3)(\sigma - 1) - 2\sigma(2\sigma + 1 - \gamma)\} - 2\sigma\gamma(T + 3)(2 - \sigma)]R, \tag{78}$$

$$+\sigma^2\gamma(T + 1)(2 - \sigma)[(2\sigma + 1 - \gamma)2(1 - \gamma) + (1 - \sigma)(T + 1) - \gamma(T + 1)(2 - \sigma)] = 0. \tag{79}$$

The loss of stability of the convection fixed points for $\sigma = 10, \gamma = -\frac{8}{3}$ using Eq. (80) is evaluated to be $R_{c2} = 25.75590$ for system parameters T = 0, R_{c2} for T = 0.1, $R_{c2} = 25.75590$ for T = 0.2, $R_{c2} = 29.344020$ for T = 0.45, and $R_{c2} = 32.775550$ for T = 0.6.

5.1 Nusselt number

According to our problem, the horizontally averaged Nusselt number for an oscillatory mode of convection is given by

$$\text{Nu}(\tau) = \frac{conduction + convection}{conduction}. \tag{80}$$

$$= \frac{\left[\frac{a}{2\pi}\int_0^{\frac{2\pi}{a_c}}\left(\frac{\partial T_b}{\partial z} + \frac{\partial T_2}{\partial z}\right)dx\right]_{z=0}}{\left[\frac{a_c}{2\pi}\int_0^{\frac{2\pi}{a_c}}\left(\frac{\partial T_b}{\partial z}\right)dx\right]_{z=0}}. \tag{81}$$

$$= 1 + \frac{\left[\frac{a}{2\pi}\int_0^{\frac{2\pi}{a_c}}\left(\frac{\partial T_2}{\partial z}\right)dx\right]_{z=0}}{\left[\frac{a_c}{2\pi}\int_0^{\frac{2\pi}{a_c}}\left(\frac{\partial T_b}{\partial z}\right)dx\right]_{z=0}}. \tag{82}$$

In the absence of the fluid motions, the Nusselt number is equal to 1. And simplifying the above equation, we will get the expressions for heat transfer coefficient:

$$\text{Nu} = 1 - 2\pi B_2(\tau). \tag{83}$$

6. Result and discussion

In this section we present some numerical simulation of the system of Eqs. (62)–(65) for the time domain $0 \leq \tau \leq 40$. The computational calculations are obtained by using Mathematica 17, fixing the values $\sigma = 10, \gamma = -8/3$, and taking in the initial conditions $X(0) = Y(0) = 0.8, Z(0) = 0.9$. In the case of T = 0, it is found that at $R_{c1} = 1$, obtained from Eq. (80), the motionless solution loses stability, and the convection solution occurs. Also the eigenvalues from Eq. (80) become equal and complex conjugate when R varies from 24.73684209 to 34.90344691 given by Gupta et al. [28]. The evolution of trajectories over a time domain in the state space for increasing the values of scaled Rayleigh number and modulation terms is given in the figures. The projections of trajectories onto Y-X, Z-Y, Z-Y, and W-Z planes are also drawn (**Figure 1**). In **Figure 2**, we observe that the trajectory moves to the steady convection points on a straight line for a Rayleigh number (R = 1:1) just above motionless solutions. It is clear from **Figure 3a** that the trajectories of the solutions approach the fixed points at R = 12, which means the motionless solution is moving around the fixed points. As the value of R changes around R = 25.75590, there is a sudden change and transition to chaotic solution (in **Figure 3b**).

In the case of gravity modulation in **Figure 4**, just keeping the values $\delta_g = 0.05, \omega_g = 10$ in connection with **Figure 3**, the motionless solution loses stability, and convection solution takes over. Even at the subcritical value of R = 25.75590, transition to chaotic behavior solution occurs, but one can develop fully chaotic nature with suitably adjusting the modulation parameter values $\delta_g = 0.05, \omega_g = 10$.

To see the effect of rotation on chaotic convection for the value of T = 0.45, we get $R_{c1} = 1.45$ from Eq. (80), which concludes that the motionless solution loses stability at this stage and the convection solution takes over. The other second and third eigenvalues become equal and complex conjugate at R = 31.44507647. In this state the convection points lose their stability and move onto the chaotic solution. The corresponding projections of trajectories and evolution of trajectories are presented in **Figure 5a** and **b**, planes Y-X, Z-X, Z-Y, and W-Z. At the subcritical value of R = 31.44507647, transition to chaotic behavior solution occurs. Observing **Figure 5b** it is clearly evident that in the presence of modulation $\sigma = 20, \delta_g = 0.2$,

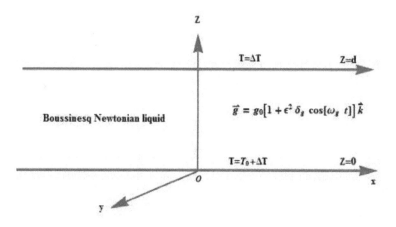

Figure 1.
Physical configuration of the problem.

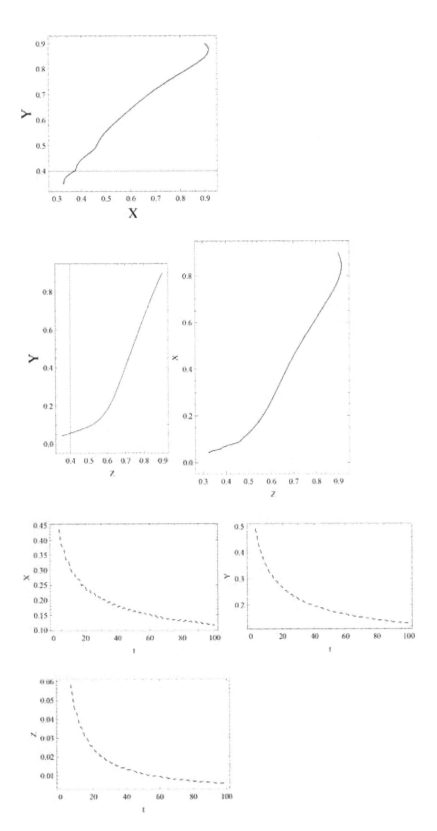

Figure 2.
Phase portraits for the evolution of trajectories over time in the state space for increasing the value of rescaled Rayleigh number (R). The graphs represent the projection of the solution data points onto Y-X, Z-X, Z-Y, and W-Z planes for $\gamma = -8/3$; $\sigma = 10$, $T = 0.1$, $R = 1.1$ $\omega_g = 0$, $\delta_g = 0.0$.

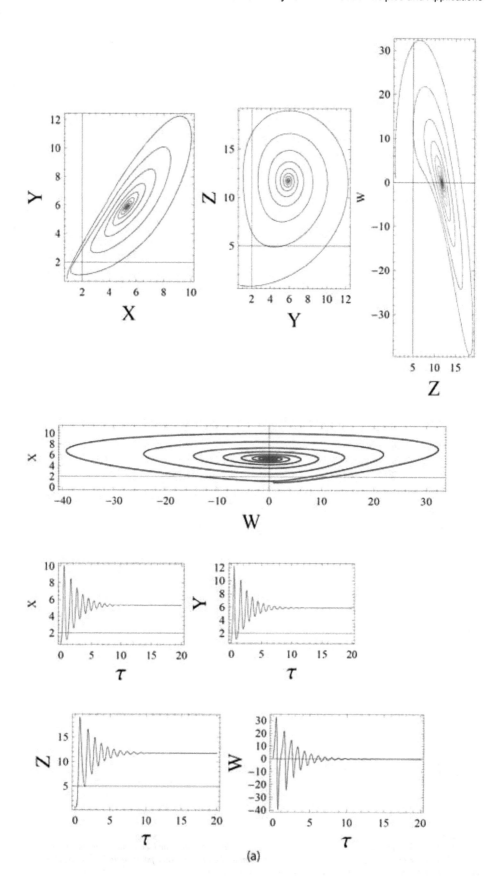

(a)

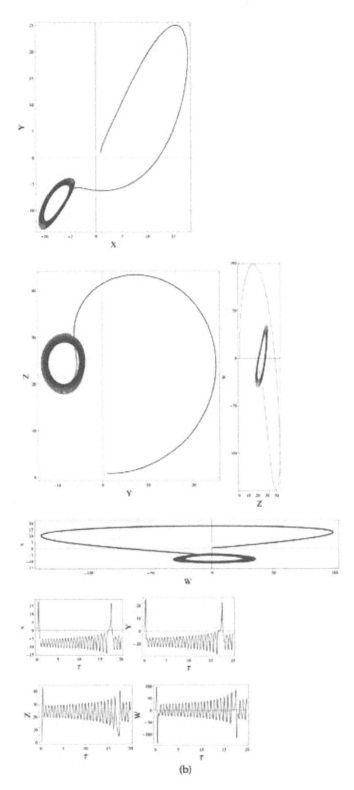

Figure 3.
(a) Phase portraits for the evolution of trajectories over time in the state space Y-X, Z-X, Z-Y, and W-Z planes for $\gamma = -8/3$, $\sigma = 10$, $T = 0.1$, $R = 12$, $\omega_g = 2$, $\delta_g = 0.0$. (b) Phase portraits for the evolution of trajectories over time in the state space Y-X, Z-X, Z-Y, and W-Z planes for $\gamma = -8/3$, $\sigma = 10$, $T = 0.1$, $R = 25.75590$, $\omega_g = 2$, $\delta_g = 0.0$.

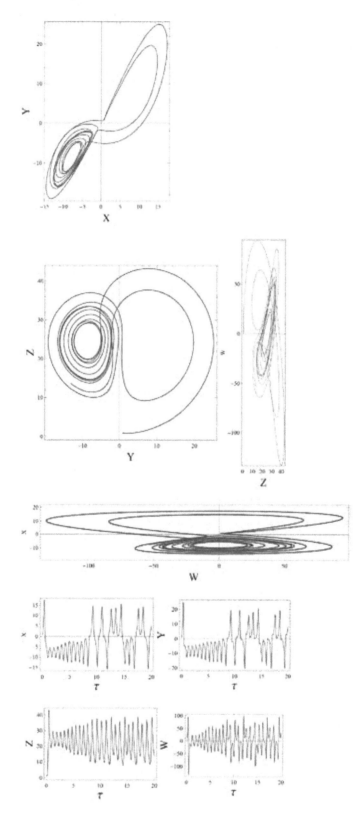

Figure 4.
Phase portraits for the evolution of trajectories over time in the state space modulation. The graphs represent the projection of the solution data points onto Y-X, Z-X, Z-Y, and W-Z planes for $\gamma = -8/3$; $\sigma = 10$, $T = 0.1$, $R = 1.1$, $\omega_g = 10$, $\delta_g = 0.05$.

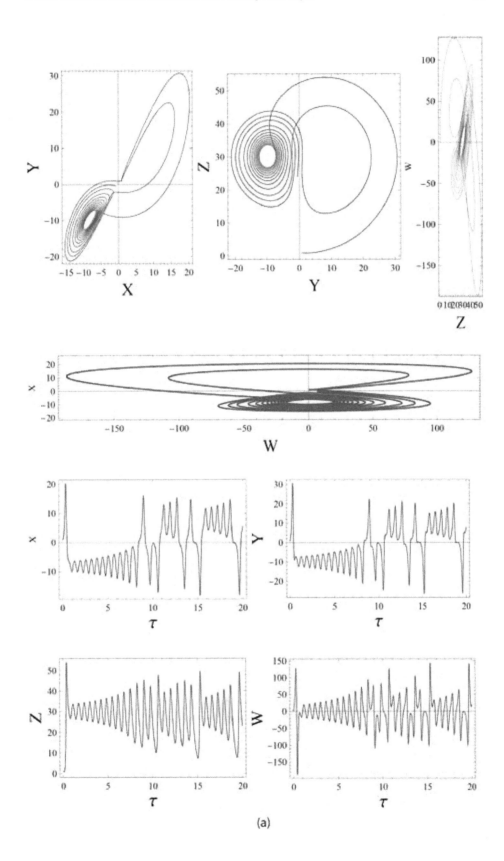

(a)

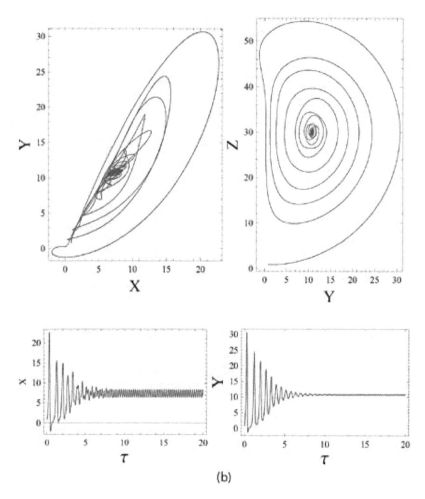

Figure 5.
(a) Phase portraits for the evolution of trajectories over time in the state space Y-X, Z-X, Z-Y, and W-Z planes for γ = −8/3, σ = 10, T = 0.2, R = 31.44507647, ω_g = 2, δ_g = 0.0. (b) Phase portraits for the evolution of trajectories over time in the state space Y-X and Z-Y planes for γ = −8/3, σ = 20, T = 0.2, R = 31.44507647, ω_g = 25, δ_g = 0.2.

$\omega_g = 20$, the trajectories are manifolds around the fixed points. Which are the interesting results to see that the system is unstable mode with rotation and buoyancy. But with gravity modulation, the system becomes stable mode.

For the value of T = 0.6, we obtain the motionless solution (where the system loss stability) given in **Figure 5b**. The values of the second and third eigenvalues become equal and complex conjugate when the value of R = 24.73684209; at this point the convection points lose their stability, and chaotic solution must occur. But due to the presence of modulation, the trend is reversed given in **Figure 6**. Observing that in the presence of modulation $\delta_g = 0.1$, $\omega_g = 2$, the system will come to stable mode for large values of R. The effect of frequency of modulation for the values $\omega_g = 2$ and $\omega_g = 20$ on chaos is presented in **Figure 7a** and **b**. It is clear that low-frequency-modulated fluid layer is in stable mode and high-frequency-modulated fluid layer in unstable mode. The reader may have look on the studies of [29–33] for the results corresponding to the modulation effect on chaotic convection.

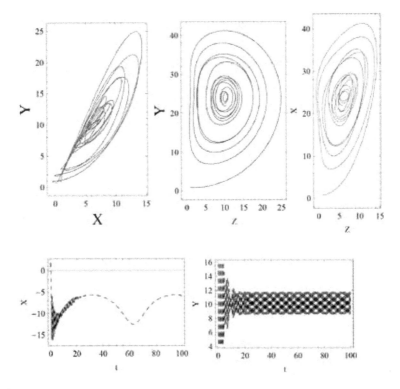

Figure 6.
Phase portraits for the evolution of trajectories over time in the state space Y-X and Z-Y planes for γ = −8/3, σ = 20, T = 0.6, R = 24.73684209, ω_g = 10, δ_g = 0.1.

Finally we also derived the heat transfer coefficient ($Nu(\tau)$) given by Eq. (83) and verified the rate of transfer of heat under the effect of gravity modulation. It is clear from **Figure 8** that heat transfer in the system is high for low-frequency modulation and for δ_g values varies from 0.1 to 0.5. The results corresponding to the gravity modulation may be observed with the studies of [19–26].

7. Conclusions

In this chapter, we have studied chaotic convection in the presence of rotation and gravity modulation in a rotating fluid layer. It is found that chaotic behavior can be controlled not only by Rayleigh or Taylor numbers but by gravity modulation. The following conclusions are made from the previous analysis:

1. The gravity modulation is to delay the chaotic convection.

2. Taking the suitable ranges of ω_g, δ_g, and R, the nonlinearity is controlled.

3. The chaos in the system are controlled by gravity modulation either from stable to unstable or unstable to stable depending on the suitable adjustment of the parameter values.

4. The results corresponding to g-jitter may be compared with Vadasz et al. [27], Kiran [31] and Bhadauria and Kiran [33].

5. It is found that heat transfer is enhanced by amplitude of modulation and reduced by frequency of modulation.

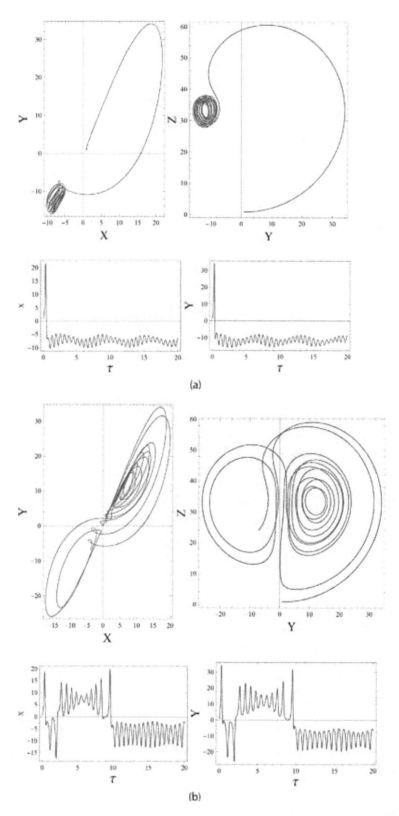

Figure 7.
(a) For $\gamma = -8/3$, $\sigma = 20$, $T = 0.2$, $R = 34.90344691$, $\omega_g = 2$, $\delta_g = 0.1$. (b) Phase portraits for the evolution of trajectories over time in the state space Y-X and Z-Y planes for $\gamma = -8/3$, $\sigma = 20$, $T = 0.2$, $R = 34.90344691$, $\omega_g = 20$, $\delta_g = 0.1$.

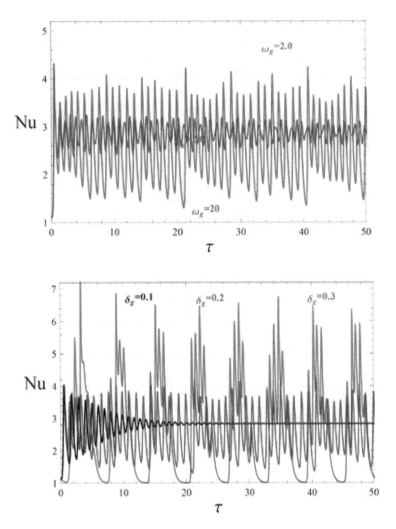

Figure 8.
Effect of ω_g and δ_g on Nu.

Acknowledgements

The author Palle Kiran is grateful to the college of CBIT for providing research specialties in the department. He also would like to thank Smt. D. Sandhya Shree (Board member of CBIT) for her encouragement towards the research. He also would like to thank the HOD Prof. Raja Reddy, Dept. of Mathematics, CBIT, for his support and encouragement. Finally the author PK is grateful to the referees for their most valuable comments that improved the chapter considerably.

Author details

Palle Kiran
Department of Mathematics, Chaitanya Bharathi Institute of Technology, Hyderabad, Telangana, India

*Address all correspondence to: pallkiran_maths@cbit.ac.in

References

[1] Lorenz EN. Deterministic non-periodic flow. Journal of Atmospheric Sciences. 1963;**20**:130-142

[2] Venezian G. Effect of modulation on the onset of thermal convection. Journal of Fluid Mechanics. 1969;**35**:243-254

[3] Gresho PM, Sani RL. The effects of gravity modulation on the stability of a heated fluid layer. Journal of Fluid Mechanics. 1970;**40**:783-806

[4] Bhadauria BS, Kiran P. Weak nonlinear oscillatory convection in a viscoelastic fluid saturated porous medium under gravity modulation. Transport in Porous Media. 2014; **104**(3):451-467

[5] Bhadauria BS, Kiran P. Weak nonlinear oscillatory convection in a viscoelastic fluid layer under gravity modulation. International Journal of Non-Linear Mechanics. 2014; **65**:133-140

[6] Donnelly RJ. Experiments on the stability of viscous flow between rotating cylinders III: Enhancement of hydrodynamic stability by modulation. Proceedings of the Royal Society of London. Series A, Mathematical and Physical Sciences. 1964;**A281**:130-139

[7] Kiran P, Bhadauria BS. Weakly nonlinear oscillatory convection in a rotating fluid layer under temperature modulation. Journal of Heat Transfer. 2016;**138**(5):051702

[8] Bhadauria BS, Suthar OP. Effect of thermal modulation on the onset of centrifugally driven convection in a vertical rotating porous layer placed far away from the axis of rotation. Journal of Porous Media. 2009;**12**(3):239-252

[9] Bhadauria BS, Kiran P. Weak nonlinear analysis of magneto–convection under magnetic field modulation. Physica Scripta. 2014; **89**(9):095209

[10] Chen GR, Ueta T. Yet another chaotic attractor. International Journal of Bifurcation and Chaos. 1999;**9**:1465-1466

[11] Vadasz P, Olek S. Weak turbulence and chaos for low Prandtl number gravity driven convection in porous media. Transport in Porous Media. 1999; **37**:69-91

[12] Vadasz P. Local and global transitions to chaos and hysteresis in a porous layer heated from below. Transport in Porous Media. 1999;**37**: 213-245

[13] Mahmud MN, Hasim I. Effect of magnetic field on chaotic convection in fluid layer heated from below. International Communications in Heat and Mass Transfer. 2011;**38**: 481-486

[14] Feki M. An adaptive feedback control of linearizable chaotic systems. Chaos, Solitons and Fractals. 2003;**15**: 883-890

[15] Yau HT, Chen CK, Chen CL. Sliding mode control of chaotic systems with uncertainties. International Journal of Bifurcation and Chaos. 2000;**10**: 113-1147

[16] Sheu LJ, Tam LM, Chen JH, Chen HK, Kuang-Tai L, Yuan K. Chaotic convection of viscoelastic fluids in porous media. Chaos, Solitons and Fractals. 2008;**37**:113-124

[17] Vadasz P. Analytical prediction of the transition to chaos in Lorenz equations. Applied Mathematics Letters. 2010;**23**:503-507

[18] Narayana M, Gaikwad SN, Sibanda P, Malge RE. Double diffusive magneto - convection in viscoelastic

fluids. International Journal of Heat and Mass Transfer. 2013;**67**:194-201

[19] Kirna P. Nonlinear throughflow and internal heating effects on vibrating porous medium. Alexandria Engineering Journal. 2016;**55**(2): 757-767

[20] Kirna P, Manjula SH, Narasimhulu Y. Oscillatory convection in a rotating fluid layer under gravity modulation. Journal of Emerging Technologies and Innovative Research. 2018;**5**(8): 227-242

[21] Kirna P, Narasimhulu Y. Centrifugally driven convection in a nanofluid saturated rotating porous medium with modulation. Journal of Nanofluids. 2017;**6**(1):01-11

[22] Kirna P, Bhadauria BS. Throughflow and rotational effects on oscillatory convection with modulated. Nonlinear Studies. 2016;**23** (3):439-455

[23] Kirna P, Narasimhulu Y. Weakly nonlinear oscillatory convection in an electrically conduction fluid layer under gravity modulation. International Journal of Applied Mathematics and Computer Science. 2017;**3**(3):1969-1983

[24] Kirna P, Bhadauria BS, Kumar V. Thermal convection in a nanofluid saturated porous medium with internal heating and gravity modulation. Journal of Nanofluids. 2016;**5**:01-12

[25] Kiran P. Throughow and g-jitter effects on binary fluid saturated porous medium. Applied Mathematics and Mechanics. 2015;**36**(10):1285-1304

[26] Vadasz JJ, Meyer JP, Govender S. Chaotic and periodic natural convection for moderate and high prandtl numbers in a porous layer subject to vibrations. Transport in Porous Media. 2014;**103**: 279-294

[27] Kiran P, Bhadauria BS. Chaotic convection in a porous medium under temperature modulation. Transp Porous Media. 2015;**107**:745-4763

[28] Gupta VK, Bhadauria BS, Hasim I, Jawdat J, Singh AK. Chaotic convection in a rotating fluid layer. Alexandria Engineering Journal. 2015;**54**:981-992

[29] Vadasz P, Olek S. Route to chaos for moderate Prandtl number convection in a porous layer heated from below. Transport in Porous Media. 2000;**41**: 211-239

[30] Kiran P, Narasimhulu Y. Internal heating and thermal modulation effects on chaotic convection in a porous medium. Journal of Nanofluids. 2018; **7**(3):544-555

[31] Kiran P. Vibrational effect on internal heated porous medium in the presence of chaos. International Journal of Petrochemical Science & Engineering. 2019;**4**(1):13-23

[32] Kirna P, Bhadauria BS. Chaotic convection in a porous medium under temperature modulation. Transport in Porous Media. 2015;**107**:745-763

[33] Bhadauria BS, Kiran P. Chaotic and oscillatory magneto-convection in a binary viscoelastic fluid under G-jitter. International Journal of Heat and Mass Transfer. 2015;**84**:610-624

3

Recent Green Advances on the Preparation of V_2O_5, ZnO and NiO Nanosheets

Daniel Likius, Ateeq Rahman, Elise Shilongo and Veikko Uahengo

Abstract

The past decade has seen a surge in the development of research on nanomaterial in the area of mixed metal oxides to fabricate ultrathin films, also known as nanosheets. In this review, different fabrication techniques of metal oxide nanosheets, such as vanadium, nickel, and zinc oxide, are presented. The chapter has also highlighted different ways of how to create smaller, affordable, lighter, and faster devices using vanadium, nickel, and zinc oxides. A detailed description of the synthesis and characterization using scanning electron microscope (SEM) and transmission electron microscope (TEM) for various shapes of nanomaterials is discussed in detail including factors that influence the orientation of nanosheets.

Keywords: nanosheets, nanoroses, nanoshapes, vanadium, nickel, zinc oxides

1. Introduction

Over the past decades, research efforts in nanoscience and nanotechnology has grown extensively globally due to the fabrication techniques of nanosheets that have attracted a great deal of research activities because of their unique physicochemi-cal properties [1]. These properties have enabled scientists to be able to design and precise control over specialized morphologies of nanomaterials [2]. Consequently, the use of nanomaterials, the tremendous potential of "nano" approaches to revolutionize the ways in which matter is synthesized, fabricated and processed is already apparent. Currently, atoms, molecules, clusters and nanoparticles are used as functional and structural units for preparing advanced and totally new materials on the nanometer length scale [3–6]. The physicochemical properties of these nanomaterials usually depend on the meticulous property employed during the fabrication process: usually by changing the dimensions of the functional and structural units of the material as well as controlling their surface morphology, it is therefore possible to tailor functionalities in exceptional ways.

Moreover, the development of this multidisciplinary field was undoubtedly accelerated by the advent of relatively recent technologies that allow the visualization, design, characterization and manipulation of nanoscale systems. Generally, a nanosheet is a two-dimensional (2D) nanostructure with thickness in a scale ranging from 1 to 100 nm. Additionally, compare to other materials such as graphene, transition metal oxide nanosheets have attracted a lot of attention recently due to their unique morphological advantages, and have shown a promising physicochemical properties for various applications. However, the fabrication of these transition metal oxide ultrafilms with controlled particles size and thickness remains a greater challenge for both fundamental study and applications. This chapter focuses on the synthesis of vanadium, nickel and zinc oxide nanosheets. The main basis for selecting these metal oxides chosen for investigation is based on the following highlights: vanadium (V), NiO, and ZnO are a highly abundant elements in the Earth's crust. Its oxides have been well known for multioxidation states (II–V) and various crystalline structures including VO_2, V_2O_5, and V_6O_{13}. Vanadium, NiO and ZnO exhibit excellent interactions with molecules or ions, outstanding catalytic activities, and/or strong electron–electron correlations [9–11].

Moreover, it also describes the morphological structures characterized by SEM and TEM techniques which include the demonstration on how these transition metal oxides change their morphology and tunable mesoporosity depend on the starting materials and heat treatment temperature.

2. Vanadium oxide nanosheets' preparation and characterization by SEM and TEM

Due to the chemical structure stability as well as the excellent physicochemical properties of vanadium oxides, these properties have attracted extensive attention for decades [7–10]. One of these vanadium oxides is vanadium pentoxide (V_2O_5) which is known as one of the best materials in nanotechnology study. Researchers have widely explored the V_2O_5 especially in alkali metal ion batteries [10]. Taking the inherent relationship between the microstructure and macroscopic properties in mind, the distinctive physicochemical properties can be synchronized and controlled via the approach of controllable synthesis of micro-/nanostructured materials. V_2O_5 nanosheets with various nanostructures such as nanosheets [11], nanoflow-ers [9], nanobelts [9], nanowires [8], nanoarrays, nanorods, nanobelts, nanonails, nanobridges, nanoprisms, nanotubes, nanobelts, nanorings, nanowhiskers, nano-combs, nanohelixes, nanosprings, nanopropeller, nanobows, nanocages, nanodisk, nanopoints, nanozigzag, nanostrings [9–11], and nanopores [8] have been fabricated. Moreover, the surface energy and surface defects of the active material are believed to contribute to the superior electrochemical performances has been developed and showed unique performances towards these specialized application. Peng et al. [11] reported V_2O_5 nanosheets with large area had been prepared via a freeze-drying process and following annealing treatment. In addition, the phase structure and morphology of V_2O_5 attained at different annealing temperatures was also systemically investigated. Xu et al. [12] reported the synthesis of $V O_{25}$ nanosheets by freeze drying method which uses hydrogen peroxide, freeze drying; post annealing and finally nanosheet can be obtained. The annealing temperature exhibited considerable influence on the microstructure of V_2O_5 nanosheets. V_2O_5 nanosheets obtained at annealing at different temperatures of 400, 450, 500 and 550°C exhibited comparatively in larger size, smaller thickness and smoother surface. **Figure 1** outlines the synthesis of V_2O_5 nanosheets as reported by Peng et al. [11].

Figure 2 shows the SEM images of V_2O_5 heat treatment at different temperature points reported by Xu et al. [12]. For V_2O_5 precursors in **Figure 2a**, layered particles are exhibited all over scattered the observed zoom. **Figure 2b** shows V_2O_5 annealed at 350°C with nanosheet microstructure with the transverse dimension greater than 10μm and the thickness of less than 20 nm. As the heat treatment temperature increased to 400°C as shown in **Figure 2c**, the transverse dimension increased while the surface morphology became smoother. The thickness of V_2O_5 nanosheet increased

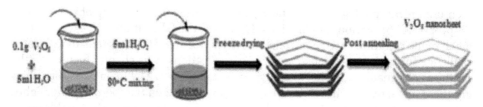

Figure 1.
Synthesis of V_2O_5 nanosheets by freeze drying method [11].

Figure 2.
SEM images of V_2O_5 nanosheet obtained at different temperature points: (a) V_2O_5 precursor, (b) 350°C, (c) 400°C, (d) 450°C, (e) 500°C, and (f) 550°C [11, 12].

by increasing the heat treatment temperature to 450°C (in **Figure 2d**) and some nanosheets gradually crumpled and agglomerated as the heat treatment temperature increased to 500°C (**Figure 2e**). The surface morphology of V_2O_5 heat treated at 550°C (**Figure 2f**) shows a mixture of agglomerated nanosheets and some regular polyhedral structures. The trend in the change of the morphological structures of V_2O_5 revealed that the heat treatment temperature possessed significant influence on the microstructure of V_2O_5. Finally, the V_2O_5 nanosheet heat treated at 400°C exhibited the optimal morphology with the largest size of transverse dimension and the thinnest thickness; hence 400°C is the optimal temperature for the fabrication of V_2O_5 nanosheet.

Liang et al. [13] reported the synthesis of vanadium nanosheets that the uniform V_2O_5 nanosheets were obtained by calcining the solvothermally prepared VO_2 nanosheets in air at 350°C for 2 h with a heating ramp of 1°C min^{-1}. Other scholars such as Cheng [14] also reported the synthesis of self-assembled V_2O_5 nanosheets/ reduced graphene oxide (RGO) hierarchical nanocomposite nanosheet also using solvothermal method. In this nanocomposite, the V_2O_5 nanosheets assembling on the RGO constitutes a 3-D hierarchical nanostructure with high specific surface area and good electronic/ionic conducting path as shown in **Figure 3** [14].

The SEM images shown in **Figure 4** were obtained to study the morphology and the structure of the as-prepared V_2O_5 nanosheets/RGO hierarchical nanostructures [14]. As shown in **Figure 4a**, large 2-D free-standing nanosheets are observed. Similarly, it can be found from close examination from **Figure 4b** that the V_2O_5 nanosheets/

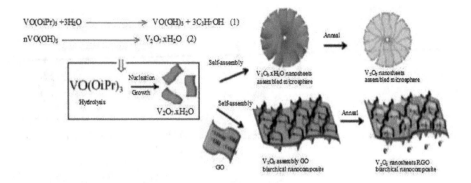

Figure 3.
Schematic illustration of the formation of 3-D V₂O₅ nanosheets/RGO hierarchical nanocomposite [14].

RGO nanocomposite is made up of multiple 2-D nanosheets on the surface of graphenenanosheets (GNS) [14]. Hence, without the addition of graphite oxide (GO), it is found that only 1-D lower-like V_2O_5 spheres consisting of many nanosheets are formed under the similar conditions. The TEM images in **Figure 5** shows that the V_2O_5 microspheres are composed of closely assembled nanosheets. Hence, by using GNS as the support, the formation of such a 3-D structure implies the effective growth of V_2O_5 nanosheets on the GNS. The structure can be further unraveled by element mapping images of carbon, vanadium and oxygen in the V_2O_5 nanosheets/RGO composite. It can be seen from **Figure 4c** that the carbon, vanadium and oxygen distributions are relatively uniform; carbon is also well dispersed all over the composites, suggesting the homogeneous dispersion of V_2O_5 on the GNS. TEM images in **Figure 5a** and **b** further show that the thin V_2O_5 nanosheets are uniformly distributed on the graphene sheets over a large area in the nanocomposite. The HR-TEM image in **Figure 5c** taken on an individual V_2O_5 nanosheet clearly shows crystal lattices with a d-spacing of 0.44 nm, corresponding to (001) planes of a crystalline orthorhombic phase of V_2O_5. The selected **Figure 5a** and **b** shows SEM images of V_2O_5 nanosheets/RGO hierarchi-cal nanocomposite at different magnifications, respectively. The insert in **Figure 4a** shows a profile of a single nanocomposite sheet; **Figure 4c** shows SEM image of V_2O_5 nanosheets/RGO hierarchical nanocomposite with corresponding EDS maps of V, O and C elements (the Au is from sputter coating) [14].

Figure 6 presents field emission scanning electron microscopy (FE-SEM) and transmission electron microscopy (TEM) images of the V_2O_5 nanosheets/CNTs nanocomposite as reported by Wang et al. [14] using the freeze drying process. It is observed that V_2O_5 nanosheets and CNTs construct a uniform and homogeneous macro-morphology **Figure 6(a)**.

The carbon nanotubes (CNTs) act as "supporting-steel-like" architectures and the V_2O_5 nanosheets are anchored on the CNTs. They combine to form a highly porous structure as shown in **Figure 6b**. The TEM images in **Figure 6d** and **e** confirms that the V_2O_5 nanosheets and CNTs form a 3D interpenetrating of network structure. It is usually recognized that most nanosheets are easily overlapped and get bunched up to form bigger bulk due to the Ostwald ripening process, leading to the decrease of actually active surface and capacity loss for electrode materials [15]. In a study reported by Chen et al. [16], V_2O_5 nanosheets are anchored on the surface of CNTs without an overlap phenomenon. As the addition ratio of CNTs increases from 0 to 20 wt%, the morphology of the V_2O_5 nanosheets/CNTs nanocomposite becomes much more homogenous and the size of the V_2O_5 nanosheets becomes much smaller. The mechanism behind the formation of the 1D and 2D nano-structures can be explained in such a way that the functional groups on the CNT

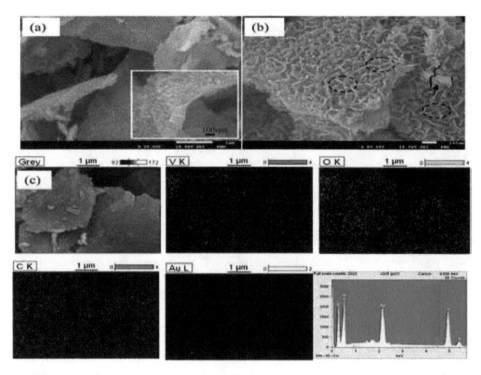

Figure 4.
(a and b) SEM images of V$_2$O$_5$ nanosheets/RGO hierarchical nanocomposite at different magnifications. The inset in (a) shows a profile of a single nanocomposite sheet; (c) SEM image of V$_2$O$_5$ nanosheets/RGO hierarchical nanocomposite with corresponding EDS maps of V, O and C elements (the Au is from sputter coating) [14].

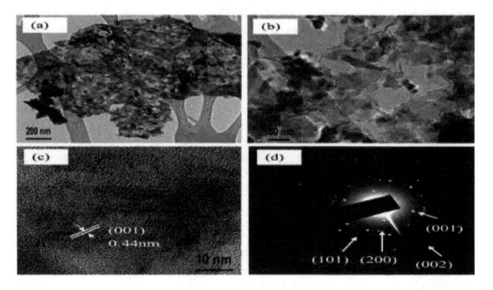

Figure 5.
(a and b) TEM images of V$_2$O$_5$ nanosheets/RGO nanocomposite at different magnifications; (c) HR-TEM image of V$_2$O$_5$ nanosheets/RGO nanocomposite; and (d) selected area electron diffraction (SAED) pattern [14].

surface act as the reaction centers. When the CNTs added into the vanadium oxide precursor solution, the vanadium oxytriisopropoxides bonded to the surface of the functional groups of CNT involved the electrostatic interaction bonding. During the freeze drying process, the hydrates in the V$_2$O$_5$ sol are frozen and removed [16].

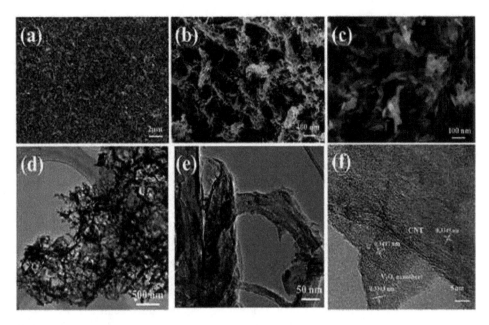

Figure 6.
The morphology observation of the V_2O_5 nanosheets/CNTs nanocomposite. (a), (b) low- and (c) high-magnification FESEM images; (d), (e) TEM and (f) HRTEM images [14].

3. Nickel oxide nanosheets preparation and characterization by SEM and TEM

Nickel (II) oxide (NiO) and nickel (II) composites in meticulously have attracted substantial interest because of a broad range of applications, namely magnetic materials [17], photovoltaic [18] Li ion batteries [19], catalysis [20], gas sensors [21], p-type transparent conducting films, infrared detectors, storage oxygen materials, fuels cells, supercapacitors, ferromagnetic oxides, gas sensors and luminescence materials, photochromic materials, [22], electrochromic windows [23], biomedicine, desalination, waste water treatment, energy related fields, catalytic reduction, adsorption, photocatalytic reduction, degradation, magnetic material, reinforcing agents in composites [24, 25]. Different methods have been reported for the synthesis of NiO, such as sol–gel [26], microemulsion [27], hydrothermal [28], co-precipitation, precipitation [29], sonochemical [15], microwave [30], metal–organic chemical vapor deposition (MOCVD), sputtering method [30], pulsed laser deposition (PLD), infrared irradiation, thermal decomposition, thermal evaporation and condensation [29, 30].

Nickel oxide is a predominantly interesting oxide because of its chemical and magnetic properties. There are various potential attractive applications of NiO in a variety of fields, such as absorbents, catalysis, battery cathodes, gas sensors, electrochromic films, magnetic materials, active optical fibers and fuel cell electrodes [31, 32]. The best know method for the fabrication of NiO nanosheets is through thermal decomposition of either nickel salts or nickel hydroxides. During this process the organic or halides on nickels burned which results in inhomogeneity of morphology and crystallite size of nickel oxide. Many efforts have been exerted to prepare NiO possessing controlled these inhomogeneity of morphology and crystallite size [33–36]. Although morphologically controlled synthesis of NiO nanocatalyst is becoming very much significant for catalytic reactions. These materials showed good electrochemical performance because of their special structure [37–41]. The fundamental process usually depends on the liquid-phase growth of ultrathin lamellar nickel hydroxide precursor under microwave irradiation as reported by Chen et al. [16]. Using urea (NH_3), during this stage, the reactions experience a homogeneous alkalinization of nickel (II) nitrate. This is followed by the hydrolysis of sodium hydroxide (NaOH) using inductive effect provided by microwave irradiation at low-temperature condition. The setup of the microwave reactor with a three-necked flask experiment is illustrated in **Figure 7a**. The optimal thermodynamic and kinetic factors are important parameters to control for the growth of ultrathin intermediate [41, 42]. The formation of nanosheets is dominated by a self-assembly and oriented attachment mechanism. The swift microwave heating allows the best saturation of reactant species for 2D anisotropic, which causes a quick formation of ultrafine nanocrystals and then spontaneous self-assembling facilitated by natural driving force of lamellar nickel hydroxide. The α-$Ni(OH)_2$ nanosheets can be totally decomposed in to NiO when annealed at 300°C (**Figure 7b-d**).

It is well noted that water molecule is a crucial reaction parameter in the fabrication of NiO nanosheet [42] (**Table 1**). In the presence of light amount of water molecules in the reactor, NiO samples are able to gain self-supporting mechanism and also exhibit a large area sheet-like morphology. However, when the amount of water decreased in the reactor, the product aggregated and became a flower-like quasi-spherical 3D hierarchical structure as shown in **Figure 8a**. It was however, observed that, they do regain their sheet-like building blocks as the water molecules removed (**Figure 8b**). In the total absence of water, the NiO surface morphol-ogy turned into spherical aggregates (**Figure 8c**) causing the disappearance of nanosheets arrangement.

Baghbanzadeh et al. [42] is in agreement with the above statements as they reported that the current methodology demonstrates that directional hydro-phobic attraction plays essential role in determining morphologies of final products. There are two factors that influence the formation of nanosheet during microwave irradiation to assist liquid-phase growth procedure. The first factor is the layered-structural nature and the second factor is the hydrophobicity. As mentioned above, the 2D anisotropic growth of nanosheets need a large driving force, thus it is possible for them to grow into nanolayer to form layered crystals. This is achieved due to the intrinsic driving force provided by lamellar $Ni(OH)_2$

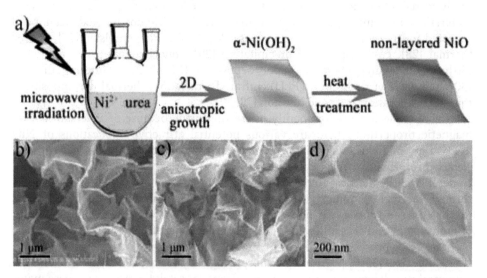

Figure 7.
(a) Schematic illustrating synthesis of nanosheets; FESEM images of (b) a-Ni(OH)$_2$ and (c), (d) NiO nanosheets [41, 42].

S. No.	Nanostructures	Particle size	Methods/conditions	Figure/ Reference
1.	V$_2$O$_5$ nanosheets	10–20 μm	Freeze drying 80°C, mixing	1–2/[1]
2.	V$_2$O$_5$ nanosheets RGO hierarchal nanocomposite	1 μm	a. Hydrothermal b. Condensation c. Nucleation calcination 350°C 2 h	3/[14]
3.	NiO nanolayered	200 nm 1 μm 20 μm	a. Microwave irradiation b. A-Ni(OH)$_2$ 2D anisotropic growth c. Heat treatment	7/[41, 42] 5/[14]
4.	NiO flower flake architecture	2 μm, 1 μm, 500 nm	SDS assisted self-nucleation assembly growth 6, 9, 12 h	10/[45]
5.	Needle-like NiO nanosheets	2 μm, 10 μm, 1 μm	a. Complexation b. Aggregation c. Fabrication	12/[46]
6.	ZnO nanosheets	200 μm, 3 μm, 500 nm	Preheating by hydrothermal method 0–24 h	14/[48] 15/[48]

Table 1.
Nanostructures of different fabrication techniques.

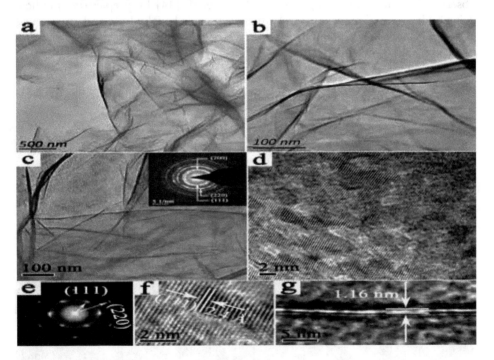

Figure 8.
(a) Low and (b) high magnification FESEM images of a-Ni(OH)$_2$ nanosheets; (c) TEM image (the inset showing SAED pattern), (d) a planar HRTEM image, (e), (f) the corresponding FFT pattern and enlarged HRTEM image recorded from (d), and (g) a vertical HRTEM image of NiO nanosheets [41, 42].

which is enough for the 2D anisotropic growth under microwave irradiation. From this, one can tell that layered-structure is a necessary requirement for the formation of 2D morphology.

As mentioned above, hydrophobicity is also one of the factors that influence the formation of nanosheet. The hydrophobicity is necessary to bring about the directional hydrophobic attraction between nanocrystals and water molecules, and it forms two phases that interface where the excessive surface energy can be accommodated [42]. There must be a balance of anisotropic hydrophobic attraction and electrostatic interaction for the spontaneous attraction of nanocrystals in order for nanosheets to be formed [43, 44]. This interaction is important to prevent their potential of shrinking and aggregating, hence allow the epitaxial orientation of the crystals. This means that the presence of the hydrophobicity terminate their stacking and packing, leading to ultrathin 2D structure rather than 3D graphite-like layered framework.

Weng et al. [45] reported the synthesis of NiO nanoflowers by hydrothermal process. **Figure 9** shows the SEM images of as-synthesized NiO. **Figure 9a** and **b** illustrates the novel hierarchical flower-like structure of the sample, which has a high similarity with the natural peony flower (inset). In **Figure 9c**, the enlarged patterns evidently exhibit that the formation of these flake-flower architectures is attributed to the partial overlapping of numerous irregular-shape ultrathin nanosheets with a thickness of approximately 10–20 nm, which are closely packed and form a multilayered structure [45].

Weng et al. [45] further reported the growth mechanism of SDS-assisted self-assembly and transformation mechanism for the synthesis of NiO flake-flower architectures. This was projected on the basis of the experimental observations and analysis reported by Zhang et al. [44] by preparation method using hydrothermal process whereby nickel chloride hexahydrate and urea served as nickel source and precipitant, respectively. The reactions with the formation proceeding are shown in Eqs. [44]:

$$CO\,(NH_2)_2 + 3H_2O \rightarrow 2NH_3\,H_2O + CO_2 \qquad (1)$$

$$NH_3\,H_2O \rightarrow NH^{4+} + OH^- \qquad (2)$$

$$CO_2 + 2OH^- \rightarrow CO^{2-}{}_3 + H_2O \qquad (3)$$

$$3Ni^{2+} + 4OH^- + CO_2{}^{3-} + 4H_2O \rightarrow Ni_3(CO_3)\,(OH)_4 \cdot 4H_2O \qquad (4)$$

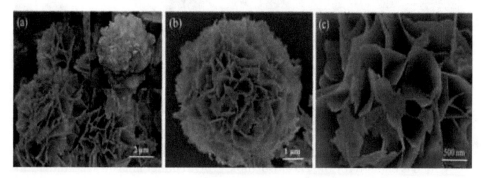

Figure 9.
(a)–(c) SEM images of the final hierarchical NiO flake-flower architectures [45].

$$Ni_3(CO_3)(OH)_4 \cdot 4H_2O \rightarrow 3NiO + CO_2 + 6H_2O \qquad (5)$$

In the initial stage, the urea in aqueous solution began to hydrolyze and release ammonia, OH^- anions as well as CO_2^{3-} anions according to Eqs. (1)–(3). The Ni^{2+}

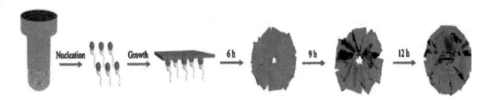

Figure 10.
Plausible SDS-assisted self-assembly and transformation mechanism for the synthesis of NiO flake-flower architectures [45].

Figure 11.
(a) and (b) SEM images of NiO samples obtained at 160°C for 6 h. (c) and (d) SEM images of NiO samples obtained at 160°C for 9 h. (e)–(f) SEM images of NiO samples obtained at 160°C for 12 h. (g) and (h) SEM images of NiO samples obtained at 160°C for 15 h [45].

species then reacted with OH⁻ and CO₂³⁻ ions to produce nickel carbonate as shown in Eq. (4). By increasing the temperature and the pH value of the solution in Eq. (5), tiny single crystals nucleated steadily due to the thermodynamic and dynamic effects [16]. With time, these tiny homogeneously nucleated crystals started to aggregate to form nanosheets through spontaneous self-organization of neighbor-ing particles aiming to have an identical crystal orientation at the planar interface, process known as oriented attachment [31]. During oriented attachment, SDS, which is an ionic surfactant, temporarily, acted also both as a structure-directing agent as well as a capping agent in the development of aggregation and enhancing absorption on the surface of the tiny crystals, respectively, in order to reduce the superficial area and their energy as illustrated in **Figure 10** [45]. However, long-chain alkyl groups from SDS produced external steric repulsion against the van der Waals attractive force so that the particles do not aggregate excessively [33, 34], which contributed to the ultra-thin nanosheets. **Figure 11** shows the SEM images of NiO nanosheets produced at the same heat treat temperature but at different annealing time.

Wen et al. [46] reported the synthesis of needle shaped NiO nanosheets by hydrothermal process as presented in **Figure 12**. The surface structures of both precursor (NiC₂O₄.2H₂O, (EG)) and final products (NiO) were characterized by FE-SEM observations. **Figure 13** shows the needle-like and flower-like structures of NiC₂O₄.2H₂O produced by varying amount of sodium oxalate [46]. The needle-like NiC₂O₄.2H₂O structures in **Figure 13a** were synthesized by using 0.061 g Sodium oxalate. By decreasing the amount of sodium oxa-late to 0.022 g, the nanoflowers of NiC₂O₄.H₂O were generated as shown in **Figure 2b** [46].

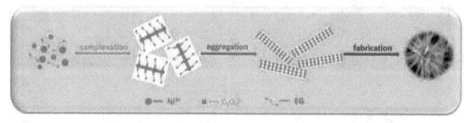

Figure 12.
Synthesis of needle-like NiO nanosheets [46].

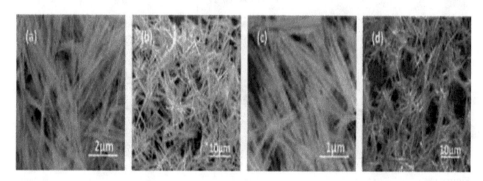

Figure 13.
SEM images of the NiC₂O₄.2H₂O and NiO: (a) and (c): needle-like architectures; (b) and (d): needle-flower architectures [46, 47].

4. Zinc oxide nanosheets preparation and morphology study by SEM and TEM

Wang et al. reported [48] the synthesis of ZnO nanostructures using a surfactant-free hydrothermal method. A schematic growth diagram of the ZnO nanostructures fabricated by preheating hydrothermal method is shown in **Figure 14**.

The morphology of the as-grown flower-like ZnO architectures were then investigated by field-emission scan electron microscopy (FE-SEM) and the results are shown in **Figure 15**.

The SEM image shown in **Figure 15(a)** demonstrates that the sample obtained after preheating for 12 h have high density flower-like ZnO architectures uniformly grow and extremely disperse in the substrates without any aggregation, with specifying high yield and good uniformity accomplished with this fabrication condition [48]. The middle magnification FE-SEM image in **Figure 15(b)** confirms that individual flowers has a diameter of about 40–50 $\mu\mu$m and consists of hundreds of thin curved nanosheets, which are spoke wise, projected from a common central zone. As presented in **Figures 15(c)** and **(d)**, high magnification FE-SEM image reveals that these ZnO nanosheets produced with the shape of flower-like architectures. The white squares in **Figure 15(e)** represent the low density ZnO nanorods which can be seen on the space without flower-like ZnO. The shape of ZnO nanorods were found to be hexagonal prism with a pyramidal top and smooth side surface and the majority of the ZnO nanorods are perpendicular to the ZnO supportive substrate. This was proved by increasing the magnification of the SEM image as shown in **Figure 15(f)**.

Wang et al. [48] also reported on the shapes of ZnO architectures removed from their supportive substrates using TEM images as shown in **Figure 16**.

The TEM images proved that the fabricated ZnO nanorods are made up of projected thin nanosheets as represented in **Figure 15a** [46, 47]. These ZnO nanosheets were found to be thin and have flat surfaces as shown in **Figure 15b**. Two sets of well resolved parallel lattice fringes are observed in high resolution TEM in **Figure 15c** [47]. The interplanar spacing is corresponding to that of {0002}

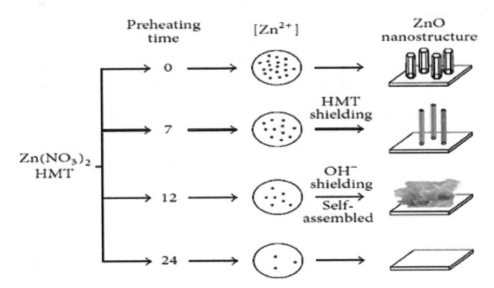

Figure 14.
Schematic growth diagram of the ZnO nanostructures fabricated by preheating hydrothermal method [48].

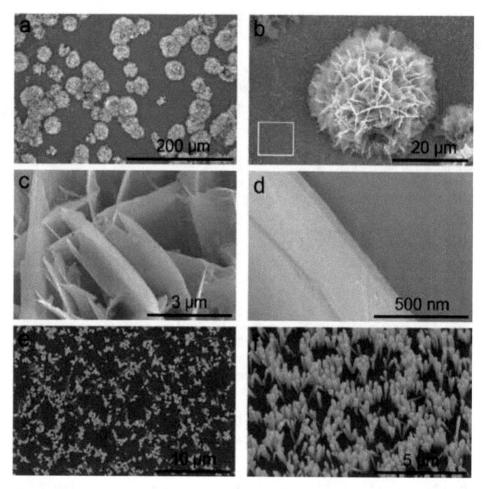

Figure 15.
FESEM images of flower-like ZnO architectures grown on ZnO thin film coated glass substrates. (a) Low magnification, (b) middle magnification, (c) and (d) high magnification of FESEM images of ZnO flowers, and (e) and (f) middle magnification of FESEM images from the space without ZnO flower, as indicated by the white square in (b) [48].

and $\{01\text{--}10\}$ planes of ZnO crystals. **Figure 15d** is its SAED pattern and exhibits visible bright spots identical to all the crystal planes of the wurtzite ZnO, indicating a single crystalline with a good crystal quality. Based on HRTEM and SAED results, Wang et al. [49] suggested that the single crystal wurtzite ZnO nanosheet grows along $[0001]$ and $[01\text{--}10]$ crystallographic directions within the $(2\text{--}1\text{--}10)$ plane. Hence, the flower-like ZnO growth process is summarized in **Figure 14**. Wang et al. [49] reported on study, Zn^{2+} and OH^- are provided by hydration of $Zn(NO_3)_2$ and HMT respectively. Therefore, the key chemical reactions can be formulated as shown in Eqs. (6)–(9):

$$(CH_2)_6N_4 + H_2O + \text{Heat} \rightarrow 4NH_{3(g)} + 6HCHO_{(g)} \qquad (6)$$

$$NH_3 + H_2O \rightarrow NH_4^+ + OH^- \qquad (7)$$

$$Zn_{2+} + 2OH^- \rightarrow Zn(OH)_{2(s)} \qquad (8)$$

$$Zn(OH)_2 \rightarrow ZnO_{(s)} + H_2O \qquad (9)$$

Wang et al. [49] have reported that in order to produce flower-like crystals of ZnO nanosheet also known as ZnO nanoroses, sodium citrate must be added in the reaction to create uniformity in the synthesis. Usually, cathodic electrodepo-sition is employed for the growth on polycrystalline ZnO thin films of different morphological orientation range from 1D (nanorods), 2D (nanoplates) to 3D crystals. However, Illy et al. [50] reported that the 2D sheet only found to occur when 1D sheets are allowed to combine under specific electrochemical condi-tions. **Figure 17** represents star-like and flower-like ZnO crystals. There is a big

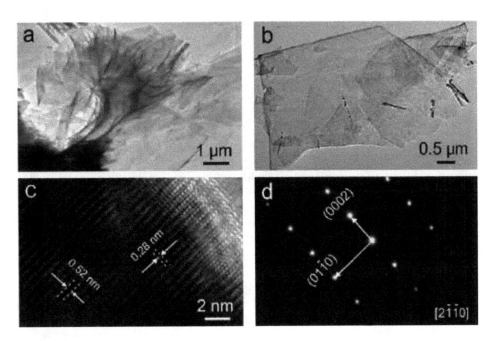

Figure 16.
TEM images of flower-like ZnO architectures. Typical TEM images of (a) an individual ZnO flower and (b) a piece of ZnO nanosheet. (c) HRTEM image and (d) SAED image of a piece of ZnO nanosheet [48].

Figure 17.
SEM images of the ZnO star-like crystals [50].

variation regarding the shape and morphology of the ZnO crystals shown. The nature and number of the reactants involved, the secondary products, concentra-tion as well as complex ions in the solution are the main factors that contribute to the kinetics of the crystal growth [50]. With regard to concentration, Illy et al. [50] demonstrated that in order to produce the ZnO stars-like crystals, the concentration of $(CH_2)_6N_4$ must be in a 1 order of magnitude higher than $Zn(NO_3)_2 \cdot 6H_2O$. In terms of secondary products as an important factor, it was found that the production of an excess of oxygen containing species in solution increased competition for the interaction with a limited number of Zinc ions. One would expect then a reduction in the size of the ZnO rods compared to those hydrothermally formed from an equimolar reactant solution. Hence, these methods of synthesizing different shapes of nanomaterials exhibit quite remark-able properties.

5. Conclusion

This review highlighted the fabrication and characterization of various shapes of metal oxide nanosheets which are characterized by SEM and TEM instru-ments. Over the past decade, there has been enormous development of research on nanosheets, in the area of metal oxides. This offers basically new capabilities to architect a broad array of novel materials in the area of metal oxides and structures on a molecular scale. This chapter summarizes different ways to create smaller, cheaper, lighter and faster devices using vanadium, nickel, and zinc oxides. Structurally controlled synthesis at large of these nanomaterials can be attained morphologically nanomaterials would have the same importance for as control-ling the helical angle of carbon nanotubes which determines it applications. These materials have remarkable applications in the all the branches of science, engineer-ing and medicine in semiconductor, electronics, biomedicine, catalysis, batteries waste water treatment, sensors, drug delivery and curing dreadful diseases and many more, which has brought about the new, developing and exciting research field called nanoscience/nanotechnology which is driving of the new millennium.

Acknowledgements

The authors acknowledge the Royal Society-DFID Africa Capacity Building Initiative for their help and support. The authors also acknowledge the support received from the Enviromental Investment Fund (EIF) of Namibia: Matching fund subsidy from National Commission on Research Science and Technology (NCRST) for strengthening capacity at universities and research institutions in Namibia.

Author details

Daniel Likius[1*], Ateeq Rahman[1*], Elise Shilongo[2] and Veikko Uahengo[1]

1 Department of Chemistry and Biochemistry, Faculty of Science, University of Namibia, Windhoek, Namibia

2 Department of Physics, Faculty of Science, University of Namibia, Windhoek, Namibia

*Address all correspondence to: daniels@unam.na and arahman@unam.na

References

[1] Li TM, Zeng W, Long HW, Wang ZC. Nanosheet-assembled hierarchical SnO$_2$ nanostructures for efficient gas-sensing applications. Actuators B. 2016;**231**:120-128

[2] Zeng Y, Wang TZ, Qiao L, Bing YF, Zou B, Zheng WTS. Synthesis and the improved sensing properties of hierarchical SnO$_2$ hollow nanosheets with mesoporous and multilayered interiors. Actuators B. 2016;**222**:354-361

[3] Eigler DM, Schweizer EK. Positioning single atoms with a scanning tunnelling microscope. Nature. 1990;**344**:524-526

[4] Heinrich AJ, Lutz CP, Gupta JA, Eigler DM. Molecule cascades. Sci-ence. 2002;**298**:1381-1387

[5] Brust M, Walker M, Bethell D, Schiffrin DJ, Whyman R. Synthesis of thiol-derivatised gold nanoparticles in a two-phase liquid–liquid system. Journal of the Chemical Society, Chemical Communications. 1994: 801-802

[6] Kiely CJ, Fink J, Brust M, Bethell D, Schiffrin DJ. Spontaneous ordering of bimodal ensembles of nanoscopic gold clusters. Nature. 1998;**396**:444-446

[7] Zhang YF, Zheng JQ , Tao H, Tian F, Meng C. Synthesis and supercapacitor electrode of VO$_2$ (B)/C core–shell composites with a pseudo-capacitance in aqueous solution. Applied Surface Science. 2016; **371**:189-195

[8] Su DZ, Zhao YJ, Yan D, Ding CH, Li JB, Jin HB. Enhanced composites of V$_2$O$_5$ nanowires decorating on graphene layers as ideal cathode materials for lithium-ion batteries. Journal of Alloys & Compounds. 2017;**695**:2974-2980

[9] Su DZ, Zhao YJ, Zhang RB, Zhao YZ, Zhou HP, Li JB, et al. Dimension meditated optic and catalytic performance over vanadium pentoxides. Applied Surface Science. 2017;**389**:112-117

[10] Wang NN, Zhang YF, Hu T. Facile hydrothermal synthesis of ultrahigh-aspect-ratio V$_2$O$_5$ nanowires for high-performance supercapacitors. Current Applied Physics. 2015;**15**:493-498

[11] Peng X, Zhang XM, Wang L, et al. Hydrogenated V$_2$O$_5$ nanosheets for superior lithium storage properties. Advanced Functional Materials. 2016;**26**:784-791

[12] Xu XM, Zhao YJ, Zhao YZ, Zhou HP, Rehman F, Li JB, et al. Self-assembly process of China rose-like β-Co(OH)$_2$ and its topotactic conversion route to Co$_3$O$_4$ with optimizable catalytic performance. CrystEngComm. 2015;**17**:8248-8256

[13] Shuquan L, Yang H, Zhiwei N, Han H, Tao C, Anqiang P, et al. Template-free synthesis of ultra-large V$_2$O$_5$ nanosheets with exceptional small thickness for high-performance lithiumion batteries. Nano Energy. 2015;**13**:58-66

[14] Cheng J, Wang B, Xin HL, Yang G, Cai H, Nie F, et al. Self-assembled V$_2$O$_5$ nanosheets/reduced graphene oxide hierarchical nanocomposite as a high-performance cathode material for lithium ion batteries. Journal of Materials Chemistry A. 2013;**1**(36): 10814-10820

[15] Yang FC, Guo ZG. Engineering NiO sensitive materials and its ultra-selective detection of benzaldehyde. Journal of Colloid and Interface Science. 2016;**467**:192-202

[16] Chen X, Li C, Grätzel M, Kostecki R, Mao SS. Nanomaterials for renewable energy production and storage. Chemical Society Reviews. 2012;**41**:7909-7937

[17] Yang M, Tao QF, Zhang XC, Tang AD, Ouyang J. Solid-state synthesis and electrochemical property of SnO_2/NiO nanomaterials. Journal of Alloys and Compounds. 2008;**459**:98-102

[18] Jiang J, Handberg ES, Liu F, Liao YT, Wang HY, Li Z, et al. Effect of doping the nitrogen into carbon nanotubes on the activity of NiO catalysts for the oxidation removal of toluene. Applied Catalysis B: Environmental. 2014:160, 716-161, 721

[19] Bai M, Dai HX, Deng JG, Liu YX, Ji KM. Porous NiO nanoflowers and nanourchins: Highly active catalysts for toluene combustion. Catalysis Communications. 2012;**27**:148-153

[20] Gu L, Xie WH, Bai SA, Liu BL, Xue S, Li Q, et al. Facile fabrication of binder-free NiO electrodes with high rate capacity for lithium-ion batteries. Applied Surface Science. 2016;**368**:298-302

[21] Zhao F, Shao Y, Zha JC, Wang HY, Yang Y, Ruan SD, et al. Large-scale preparation of crinkly NiO layers as anode materials for lithium-ion batteries. Ceramics International. 2016;**42**:3479-3484

[22] Cui F, Wang C, Wu SJ, Liu G, Zhang FF, Wang TM. Lotus-root-like NiO nanosheets and flower-like NiO microspheres: Synthesis and magnetic properties. CrystEngComm. 2011;**13**:4930-4934

[23] Zhu P, Xiao HM, Liu XM, Fu SY. Template-free synthesis and characterization of novel 3D urchin-like α-Fe_2O_3 superstructures. Journal of Materials Chemistry. 2006;**16**:1794-1797

[24] Wang W, Zeng ZCW. Assembly of 2D nanosheets into 3D flower-like NiO: Synthesis and the influence of petal thickness on gas-sensing properties. Ceramics International. 2016;**42**:4567-4573

[25] Tian K, Wang XX, Li HY, Nadimicherla R, Guo X. Lotus pollen derived 3-dimensional hierarchically porous NiO microspheres for NO_2 gas sensing. Sensors and Actuators B: Chemical. 2016;**227**:554-560

[26] Lin LY, Liu TM, Miao B, Zeng W. Hydrothermal fabrication of uniform hexagonal NiO nanosheets: Structure growth and response. Materials Letters. 2013;**102-103**:43-46

[27] Lu Y, Ma YH, Ma SY, Jin WX, Yan SH, Xu XL, et al. Synthesis of cactus-like NiO nanostructure and their gas-sensing properties. Materials Letters. 2016;**164**:48-51

[28] Miao B, Zeng W, Lin LY, Xu S. Hydrothermal synthesis of nano sheets. Physica E. 2013;**52**:40-45

[29] Kumar R, Baratto C, Faglia G, Sberveglieri G, Bontempi E, Borgese L. Tailoring the textured surface of porous nanostructured NiO thin films for the detection of pollutant gases. Thin Solid Films. 2015;**583**:233-238

[30] Sta I, Jlassi M, Kandyla M, Hajji M, Koralli P, Krout F, et al. Surface functionalization of sol–gel grown NiO thin films with palladium nanoparticles for hydrogen sensing. International Journal of Hydrogen Energy. 2016;**41**:3291-3298

[31] Carnes C-L, Klabunde K-J. The catalytic methanol synthesis over nanoparticle metal oxide catalysts. Journal of Molecular Catalysis A. 2003;**194**:227-236

[32] Ralston D, Govek M, Graf C, Jones D-S, Tanabe K, Klabunde K-J, et al. Fuel processing nano particles. Fuel Processing Technology. 1978;**1**:143-146

[33] Pradhan B-K, Kyotani T, Tomita A. Nickel nanowires of 4 nm diameter in the cavity of carbon

nanotubes. Chemical Communications. 1999:1317-1318

[34] Park J et al. Monodisperse nanoparticles of Ni and NiO: Synthesis, characterization, self-assembled superlattices, and catalytic applications in the Suzuki coupling reaction. Advanced Materials. 2005; **17**:429-434

[35] Yan H, Blan C-F, Holland B-T, Parent M, Smyrl W-H, Stein A. A chemical synthesis of periodic macroporous NiO & metallic Ni. Advanced Materials. 1999;**11**:1 003-1006

[36] Wang Y, Zhu Q, Zhang H. Carbon nanotube-promoted Co–Cu catalyst for highly efficient synthesis of higher alcohols from syngas. Chemical Communications. 2005:523-525

[37] Pang H, Lu Q, Li Y, Gao F. Facile synthesis of nickel oxide nanotubes and their antibacterial, electrochemical and magnetic properties. Chemical Communications. 2009: 7542-7544

[38] Yang Q, Sha J, Ma X, Yang D. Synthesis of NiO nano wires by a sol gel process. Materials Letters. 2005; **59**:1967-1970

[39] Wang W, Liu Y, Xu C, Zheng C, Wang G. Synthesis of NiO nanorods by a novel simple precursor thermal decomposition approach. Chemical Physics Letters. 2002;**362**:119-122

[40] Song LX, Yang ZK, Teng Y, Xia J, Du P. Nickel oxide nanoflowers: Formation, structure, magnetic property and adsorptive performance towards organic dyes and heavy metal ions. Journal of Materials Chemistry A. 2013;**1**:8731-8736

[41] Baghbanzadeh M, Carbone L, Cozzoli PD, Kappe CO. Microwave-assisted synthesis of colloidal inorganic nanocrystals. Angewandte Chemie, International Edition. 2011;**50**:11312-11359

[42] Bilecka I, Niederberger M. Microwave chemistry for inorganic nanomaterials synthesis. Nanoscale. 2010;**2**:1358-1374

[43] Tang Z, Zhang Z, Wang Y, Glotzer SC, Kotov NA. Self-assembly of CdTe nanocrystals into free-floating sheets. Science. 2006;**314**:274-278

[44] Zhang Z, Tang Z, Kotov NAS, Glotzer C. Spontaneous CdTe alloy CdS transition of stabilizer-depleted CdTe nanoparticles induced by EDTA. Nano Letters. 2007;**7**:1670-1675

[45] Zeng W, Miao R, Gao Q. SDS-assisted hydrothermal synthesis of NiO flake-flower architectures with enhanced gas-sensing properties. Applied Surface Science. 2016;**384**:304-310

[46] Zeng W, Cao S, Long H, Zhang H. Hydrothermal synthesis of novel flower-needle NiO architectures: Structure, growth and gas response. Materials Letters. 2015;**159**:385-388

[47] Qiu J, Weng B, Zhao L, Chang C, Shi Z, Li X, et al. Nanoflakes NiO synthesis. Journal of Nanomaterials. 2014;**11**. Art I281461

[48] Wang ZL, Kong XY, Ding Y, Gao PX, Hughes WL, Yang RS, et al. Semiconducting and piezoelectric oxide nanostructures induced by polar surfaces. Advanced Functional Materials. 2004;**14**:943-956

[49] Illy B, Shollock BA, MacManus-Driscoll JL, Ryan MP. Electrochemical growth of ZnO nanoplates. Nanotechnology. 2005;**16**:320-324

[50] Palumbo M, Henley SJ, Lutz T, Stolojan V, Silva SRP. A fast sonochemical approach for the synthesis of solution processable ZnO rods. Journal of Applied Physics. 2008;**104**:074906-0749066

A Compact Source of Terahertz Radiation Based on an Open Corrugated Waveguide

Ljudmila Shchurova and Vladimir Namiot

Abstract

We show that it is possible to produce terahertz wave generation in an open waveguide, which includes a multilayer dielectric plate. The plate consists of two dielectric layers with a corrugated interface. Near the interface, there is a thin semiconductor layer (quantum well), which is an electron-conducting channel. The generation and amplification of terahertz waves occur due to the efficient energy exchange between electrons, drifting in the quantum well, and the electromagnetic wave of the waveguide. We calculate the inhomogeneous electric fields induced near the corrugated dielectric interface by electric field of fundamental mode in the open waveguide. We formulate hydrodynamic equations and obtain analytical solutions for density waves of electrons interacting with the inhomogeneous electric field of the corrugation. According to numerical estimates, for a structure with a plate of quartz and sapphire layers and silicon-conducting channel, it is possible to generate electromagnetic waves with an output power of 25 mW at a frequency of 1 THz.

Keywords: terahertz source, corrugated waveguide, open waveguide, drifting electrons, interaction of electrons with a wave

1. Introduction

For past few decades terahertz radiation, which occupies the bandwidth from approximately 0.3–10 THz, have received a great deal of attention. Devices exploiting this waveband are set to become increasingly important in a very diverse range of applications, including medicine and biology [1, 2]. Nevertheless, despite significant progress in the study of terahertz sources in recent years (see, for example, [3, 4] and references therein), this range is mastered much less than its neighboring frequency ranges: the optical range, in which optoelectronic devices are used, and the microwave range, in which electro-vacuum microwave devices are mainly used.

Currently, the key problem lies in the creation of a sufficiently intense, and at the same time compact, terahertz source that can be adapted for a variety of applications.

Here we consider a scheme of a compact terahertz generator, which uses some methods and ideas that have been successfully applied in vacuum microwave generators. In the schemes of terahertz generator discussed here, as in microwave generators (such as backward wave and traveling wave tubes), electrons interact with the waveguide wave and transfer their energy to this wave. However, a straight-forward adaptation of microwave generator circuit for terahertz generators causes serious difficulties. It is known that the most efficient energy transfer between electrons and an electromagnetic wave occurs when the electron velocity is close to the phase velocity of the wave. However, since the electron velocity is much less than the speed of the electromagnetic wave in a vacuum (light speed), slow-wave structures are used to reduce the wave velocity. The characteristic dimensions of such slow-wave structures should be comparable with the length of the amplified electromagnetic wave. There is a strong absorption of electromagnetic radiation in slow-wave structures with small characteristic dimensions, which are necessary for slowing down terahertz waves. This significantly affects the generation conditions. It is extremely difficult both to slow the wave and to avoid losses in this frequency range.

Here, we offer another phase-matching method. In the proposed terahertz generator, a speed of the electromagnetic wave, which propagate along a waveguide, is close to the speed of light in vacuum, but electrons are still able to effectively transfer their energy to the wave. Such a situation may occur in corrugated waveguides, in which the corrugated dielectric surface is located in the region of a high electric field of an electromagnetic wave, and electrons move over this corrugated surface. The electromagnetic wave causes the polarization of the dielectric in the zone of the corrugation, and this polarization, in turn, induces an alternating electric field near the corrugated dielectric interface. The characteristic scale that determines the induced electric field is determined by the corrugation period. Since the corrugated interface has a periodic structure, the induced electric field near the interface can be regarded as the sum of an infinite set of harmonics. By selecting the size of the period, we can ensure that electrons, moving at a small distance above the corrugation, would effectively interact with the first harmonic of the induced electric field and would give its energy to the wave.

Here we consider a scheme of an open corrugated waveguide, which includes a thin multilayer dielectric plate (**Figure 1**). The plate consists of two different dielectric layers with a dielectric interface corrugated periodically in the transverse direction of the waveguide (in the direction perpendicular to the direction of wave propagation). The electric field of the transverse electromagnetic wave (TE wave) induces near the corrugation an electric field that is non-uniform in the transverse direction of the waveguide.

There is a quasi-two-dimensional conducting channel (quantum well) in one of the dielectric layers. In the quantum well located near the corrugation, electrons drift in the transverse direction of the waveguide, and interact with the transverse electric field of the induced surface electromagnetic wave. Phase matching ($v_e \approx v_f$) can be achieved by selecting the applied voltage (which determines the electron drift velocity v_e) and the corrugation period L (which defines wave number $k_c = 2\pi/L$ and the phase velocity of the surface wave $v_f = \omega/k_c$, where ω is the frequency of the waves). The phase velocity of electromagnetic waves with a frequency of $\sim 10^{12} s^{-1}$ is a value of the order of $\sim 10^6 - 10^7$ cm/s for the corrugation period of $L \sim 0.1-1$ µm (structures with such parameters are achievable at the present technological level). The drift velocity of electrons in a conducting channel can have a value of the order of $\sim 10^6 - 10^7$ cm/s. Thus, the corrugation period and the parameters of the electronic system can be selected so as to satisfy the conditions of the most effective interaction of the carriers with the electromagnetic wave. As a result of the interaction, the amplitude of the electric field of the

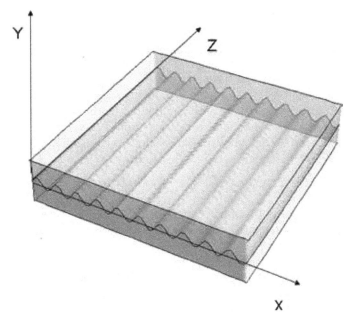

Figure 1.
A plate of two different dielectric layers with a corrugated dielectric interface.

transverse electromagnetic wave increases, while the amplified electromagnetic wave itself propagates along the waveguide at a speed close to the speed of light in vacuum $(c >> v_f)$.

We propose a generator circuit of an open waveguide, which includes a thin multilayer dielectric plate. In such waveguide, the field of an electromagnetic wave is focused in a region that includes both the plate itself and some region near the plate. Ohmic losses of electromagnetic waves are only into the dielectric plate, while wave energy is concentrated mainly outside the plate. So, in open waveguides, it is possible to reduce the energy loss of an electromagnetic wave (and, thus, to improve the generation conditions).

In our works [5, 6], a similar synchronization scheme was proposed for a open corrugated waveguide, in which electrons move ballistically in vacuum above a corrugated plate surface and interact with the non-uniform electric field induced near the corrugation. Such electro-vacuum terahertz generators can have an output power of the order of watts, and efficiency up to 80% [5].

However, in such schemes it is necessary to stabilize the electron beam position above the plate, avoiding the bombardment of the plate by electrons. In [5], we proposed a method for stabilizing the electron beam, but this greatly complicated the scheme of the terahertz generator.

In the scheme considered here, the position of the electrons is already stabilized, since the electrons drift in the quantum well, and the conducting channel can be located close to the corrugated dielectric interface. Such a terahertz generator is compact and, moreover, quite simple to produce. However, the parameters of such a generator are low compared with the parameters of the generator described in [5]. In the terahertz generator scheme considered here, electrons drift in a semiconductor quantum well and experience a large number of collisions. As a result, there are significant losses in Joule heat in such a waveguide. According to our estimates, in such a scheme, it is possible to generate terahertz waves with an output power of tens and hundreds of milliwatts and efficiency of the order of 1%.

A brief description of the terahertz generator, based on the interaction of electrons in a quantum well with an electromagnetic wave of a corrugated waveguide, was previously presented in our works [7, 8].

In the second section of this chapter, we present calculations of the induced inhomogeneous electric field of the wave near the corrugated dielectric interface. In the third section, we use a hydrodynamic approach to describe a system of charged carriers, which drift in a quantum well and interact with the inhomogeneous electric field of the wave in the zone of the corrugation. We define the parameters of the system under which the amplification of the electromagnetic field is the most effective. In the fourth section, we give a brief conclusion.

2. Wave electric fields in an open waveguide with a plate consisting of two dielectric layers with a corrugated interface

The generator circuit considered here is an open waveguide, which includes a thin dielectric plate consisting of two dielectric layers with a dielectric interface, periodically corrugated in the transverse direction of the waveguide. An electron conducting channel is included in one of the dielectric layers of the plate close to the corrugated dielectric interface (**Figure 2**), and the electrons drift in the conducting channel in the transverse direction of the waveguide (in the direction perpendicular to the direction of wave propagation).

We assume that the conducting channel is sufficiently thin (its thickness is much smaller than the corrugation amplitude). In this case, the problem of calculating of electric fields in the waveguide, and the problem of calculating of interactions of electrons with electromagnetic waves can be solved separately. In this section, we solve the problem of calculating of electric fields without electrons. In the next section, we introduce electrons, which interact with wave electric fields, into the picture.

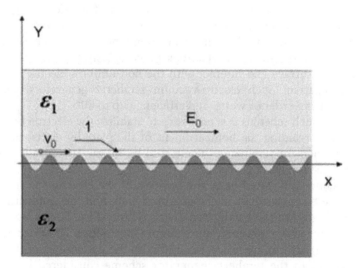

Figure 2.
Plate profile of an open waveguide. The plate consists of two dielectric layers with permittivities ε_1 (light gray) and ε_2 (dark gray) having a periodically corrugated dielectric interface. Electrons drift with a mean velocity v_0 in a semiconductor layer 1. E_0 is an electric field of a principal wave of the waveguide (TE-wave).

Electromagnetic waves, including waves in the terahertz frequency range, can propagate in the proposed open waveguide. The waveguide contains a dielectric plate of thickness $2a$ consisting of two dielectric layers with different permittivities ε_1 and ε_2 ($\varepsilon_1 < \varepsilon_2$). For sufficiently thin plate, $a << c/\omega$, and for not too large values of ε_1 and ε_2 ($\varepsilon_i < 10$) the electromagnetic wave energy is mostly concentrated in the region outside the plate. In the waveguide, an electromagnetic wave propagates with the velocity close to the speed of light.

Let us consider the case of a TE wave, in which there is an antinode of the electric field in the center of the cross section of the plate (where the dielectric plate is) [7] (**Figure 1**).

For a thin plate ($a << 2\pi c/\omega$) with a flat dielectric interface, the electric field of the TE wave is directed along the X-axis, and is given by

$$E_{TE}(r,t) = e_x E_x(y) \cdot \exp{(ikz - i\omega t)}, \tag{1}$$

The field E_x can be represented as:

$$E_x \approx E_{in} \cos{(\chi y)}, \; if \; |y| \leq a, and \\ E_x \approx E_{out} \exp{(-s|y|)}, \; if \; |y| \geq a \tag{2}$$

Here, e_x is a unit vector along the x-axis, $E_{out} \approx E_{in} = E_0$, χ and s are transverse components of the wave vector in a dielectric and vacuum, respectively. The values χ and s are related by $sa = \frac{1}{\varepsilon} \cdot \chi a \cdot tg \; (\chi a)$ and $(\chi a)^2 + (sa)^2 = \frac{\omega^2 a^2}{c^2}(\varepsilon - 1)$, where $\varepsilon = \sqrt{\varepsilon_1 \cdot \varepsilon_2}$ is the average value of permittivity of the dielectric plate.

However, the electric field E_x of the TE-wave is uniform for a homogeneous dielectric plate, as well as multi-layer plates with a flat dielectric interface. In this case, the effective interaction between electron and the wave should be absent.

We propose a scheme of a waveguide with a thin multilayered dielectric plate, and the interface between two dielectric layers is a corrugated surface (**Figure 1**). The waveguide with this plate can also serve as a waveguide for the TE wave, and the wave propagation speed along the Z-axis is still being close to the speed of light. But in this case, in the vicinity of corrugated dielectric interface, an additional surface wave is induced in a field of the TE wave. The induced inhomogeneous electric field of the surface wave is comparable with the electric field of the volume electromagnetic wave only in a very narrow region near the dielectric interface. And the inhomogeneous electric field decreases exponentially (in Y-direction) with distance from the interface. The induced electric field is inhomogeneous in the Z-direction of the waveguide, and electrons also drift in X-direction in quasi-two-dimensional conductance channel, located in the zone of the corrugation (**Figure 2**). Such electrons can interact with an inhomogeneous field of the wave.

Let the coordinates of the dielectric interface in the waveguide cross section vary according to a periodic law, for example, to the law $y(x) = R \cos{(k_c x)}$. Here $k_c = 2\pi/L$, and the corrugation amplitude R is significantly smaller than the plate thickness, $R << 2a$, and the corrugation period L is much less than the plate width.

We accept the condition

$$k_c >> k, \tag{3}$$

and, as will be clear from the following, it is worth considering the case $k_c R \sim 1$, in which the interaction of an electromagnetic wave with electrons (and the amplification of this wave) will be most effective. We assume that the conducting channel is thin enough ($h < < R$, where h is the channel width), so that the electrons in the channel do not deform the electromagnetic fields in the dielectric.

In such waveguide, the electric field of the wave can be represented as E $(r,$ $t) = E_{TE}$ $(r,$ $t) + F$ $(r,$ $t)$, where E_{TE} $(r,$ $t)$ is given by (1) and (2) [7]. The electric field F $(r,$ $t)$ of the wave, which is induced near the corrugation, is comparable with the TE wave field only in a very narrow region (with a width of about R) near the dielectric interface. And F $(r,$ $t)$ is significantly less than E_{TE} $(r,$ $t)$ outside this region.

Let us proceed to calculation of the electric field F $(r,$ $t)$. The field F $(r,$ $t)$ is caused by influence of the electric field E_{TE} $(r,$ $t)$ of the TE-wave on the dielectric interface. However, due to the smallness of the corrugation size is compared to the electromagnetic wavelength $\lambda = 2\pi/\chi$ ($\lambda > > R$, $\lambda > > L$), the TE wave's electric field can be considered as a potential at distances typical for corrugation. Let us denote this field by E_0, where $E_0 \cos(\chi y) \approx E_0$. Since the corrugated dielectric interface is homogeneous along the Z-axis, the problem of calculating F $(r,$ $t)$ is effectively two-dimensional one [7].

The problem can be described by the scalar potential $\phi^{(i)}$, which satisfies to the Laplace equation. We denote the potential as $\phi^{(1)}(x, y)$ inside the dielectric layer with the permittivity ε_1, and as $\phi^{(2)}(x, y)$ inside the layer with the permittivity ε_2 [7]. The functions $\phi^{(1)}(x, y)$ and $\phi^{(2)}$ (x, y) satisfy to the Laplace equation $\Delta\phi^{(1)}(x, y) = 0$ and $\Delta\phi^{(2)}(x, y) = 0$ within its domain. Then the boundary conditions take the form

$$\phi^{(1)}(x, y)\big|_c = \phi^{(2)}(x, y)\big|_c \tag{4}$$

$$\varepsilon_1 \frac{\partial\phi^{(1)}(x, y)}{\partial n}\bigg|_c = \varepsilon_2 \frac{\partial\phi^{(2)}(x, y)}{\partial n}\bigg|_c. \tag{5}$$

The conditions (4)–(5) represent the continuity of the tangential electric field and normal component of an electric induction vector in a point with coordinates (x, y) on the corrugation dielectric interface [7].

Since the corrugated dielectric interface is described by a periodic function of x, the electrostatic potential functions $\phi^{(i)}(x, y)$ can be written as the sum of spatial harmonics:

$$\phi^{(i)}(x, y) = \sum_{m=0}^{\infty} B_m^{(i)} \cos(mk_c x) \exp(-mk_c|y|)$$
$$+ \sum_{m=0}^{\infty} A_m^{(i)} \sin(mk_c x) \exp(-mk_c|y|) + E_0 x \tag{6}$$

In our problem, $y(x) = R$ $\cos$ $(k_c x)$ is a continuously differentiable function. In this case, expressions (6) represent a complete set of basis functions [9] for potentials $\phi^{(i)}$ and for electric fields $E^{(i)} = -\nabla\phi^{(i)}$. The expansion coefficients $A_m^{(i)}$ and $B_m^{(i)}$ of the sum (6) are found from the boundary conditions (4)–(5) at the point $(x, y) = (x, \cos$ $k_c x)$ [7]. For the boundary described by $y(x) = R$ $\cos$ $(k_c x)$, this system is consistent, if the coefficients $B_m^{(i)} = 0$, and $A_m^{(i)} \neq 0$. Then for the calculation of coefficients $A_m^{(i)}$ we have limited the harmonic expansion to the seventh term. According to our calculations, when $\varepsilon_2/\varepsilon_1 \geq 3$, taking into account the seven expansion terms is sufficient for calculating the coefficients, and the error will not exceed a few percent.

The resulting expression for the x-component $E_x^{(1)}$ of the electric field $E^{(1)} = -\nabla\phi^{(1)}$ induced in the upper layer near the corrugation (at $y > R\ \cos\ (k_c x)$) can be represented as

$$E_x^{(1)}(x, y) = E_0 \sum_{m=1}^{\infty} C_m^{(1)} e^{imk_c x} e^{-mk_c|y|} \qquad (7)$$

The coefficients $C_m^{(1)}$, which characterize the contribution of mth harmonic to the electric field induced above the corrugation, depend on the shape and size of corrugation and, moreover, on the parameters ε_1 and ε_2. We carried out numerical analysis on an example of a structure with a plate of quartz (ε_1 = 4.5) and sapphire (ε_2 = 9.3). In the case, when the dielectric boundary is described by $y(x) = R\ \cos\ (k_c x)$, each of the coefficients $C_m^{(1)}$ can be represented as a function of $k_c R$. **Figure 3** shows the calculated dependences $C_1^{(1)}(k_c R)$, $C_2^{(1)}(k_c R)$ and $C_3^{(1)}(k_c R)$. For $k_c R \geq 1$, the value of $C_1^{(1)}$ is much greater than $C_2^{(1)}$ and $C_3^{(1)}$. The coefficients $C_m^{(1)}$ for $m > 3$ are much less than $C_1^{(1)}$, $C_2^{(1)}$ and $C_3^{(1)}$.

Figure 4 shows the first harmonic of electric field $E_x^{(1)}/E_0$ at $y = R$ as a function of $k_c x$ for different values of the corrugation amplitude ($k_c R$ = 0.3, 1.2, 2). At $y = R$, the amplitude of $E_x^{(1)}/E_0$ takes the maximum value for $k_c R \approx$ 1.2. At the maximum we have $E_x^{(1)}/E_0$=0.15 and $C_1^{(1)} \approx 0.5$. At the corrugation height $R \approx 1.2 \cdot (k_c)^{-1}$ the amplitude of the first harmonic $E_x^{(1)}$ of the electric field reaches its maximum value. The first harmonic electric field of the corrugation is smaller for both the higher and lower corrugation amplitude. Indeed, for lower corrugation heights ($k_c R << 1$), the induced electric field is small because of weak polarization of charges on the dielectric interface with weak corrugation. In the case of higher corrugation heights ($k_c R >> 1$), the charge-induced field largely decays at $y \approx R$.

Thus, we have the expression for the x-component of the first harmonic electric field of a wave at $y = R$ (in the region where the conducting channel is)

$$F_x^{(1)} = E_0 C_1^{(1)} \exp\ (-k_c R) \exp\ (ik_c x + ikz - i\omega t). \qquad (8)$$

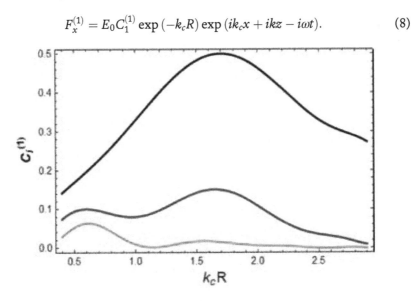

Figure 3.
Coefficients, characterizing a contribution of the first $C_1^{(1)}$ (black line), second $C_2^{(1)}$ (blue line) and third $C_3^{(1)}$ (green line) harmonics to the induced electric near the corrugation, as a function $k_c R$; R is a corrugation amplitude, $k_c = 2\pi/L$, L is a corrugation period.

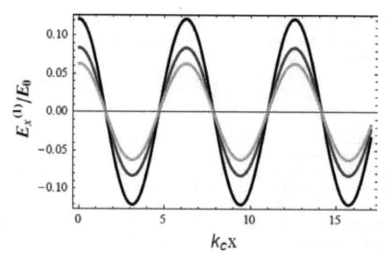

Figure 4.
Calculated values of the first harmonic electric field at $y = R$ as a function of normalized coordinate $k_c x$ for different values of a corrugation amplitude: $R = 0.3/k_c$ (blue line), $R = 1.2/k_c$ (black line) and $R = 2/k_c$ (green line).

The optimal corrugation amplitude R and period L, in which the first harmonic amplitude of the non-uniform electric field takes its maximum, are determined by $k_c R \approx 1.2$ (where $k_c = 2\pi/L$).

3. Electron interaction with electromagnetic wave

Let us now consider the problem of interaction of an electromagnetic wave with the electrons. In our study frame electrons drift in a quasi-two-dimensional conducting channel in the transverse direction (X direction) of the waveguide in a quasi-two-dimensional conducting channel (**Figure 2**). The conductive channel is located close to the corrugated dielectric interface, that is, in large inhomogeneous electric field region.

The most effective interaction between the first harmonic of a wave and electrons happens under the condition of synchronism, when phase velocity $v_f = \omega/k_c$ of the wave is close to the electron drift velocity v_0. The corrugation period L and applied voltage are selected in such a way as to satisfy the condition $v_0 \approx v_f$. (More precisely, the electron drift velocity must slightly exceed the phase velocity of the electromagnetic wave [7].) Due to the change of the electron velocity in the electric field, the electron flow becomes velocity modulated. Electron velocity modulation process is accompanied an electron bunching, and electron density wave is formed. Parts of the electrons, which decelerated in the electric field of the corrugation, release their energy to the electromagnetic wave. The slightly amplified electromagnetic wave causes an increase in the amplitude of the electron density wave which, in turn, amplifies the electromagnetic wave even more [5]. Thus, an oscillatory mode appears.

To describe the electrons interacting resonantly with the wave, a self-consistent system of equations is required. The system should take into account the change as electromagnetic wave fields, and electron velocity $v(x,t)$ and density $n(x,t)$. We assume that the electron density is quite low and the conducting channel is very thin, so that the screening of the corrugation electric field by electrons is negligible. TE-wave of the waveguide includes H_y and H_z components of the magnetic field. Since the electron collision frequency is much greater than the cyclotron frequency, the influence of magnetic field on the electron motion is negligible. The form of a solution describing the electric field of the first harmonic wave is described by (8). Since the corrugated dielectric interface is homogeneous along the Z-axis we can take $z = 0$ without loss of generality. At the initial generation stage, while the linear approximation is correct, the amplitude of the electromagnetic increases with time by the exponential law with an increment γ. Then, taking into account (8), the contribution of the first harmonic to the wave's electric field at $y = R$ can be presented as

$$F_x^{(1)}(x,t) = E_0 \cdot C_1^{(1)}(k_c R) \exp\left(-k_c R\right) \cdot \exp\left(ik_c x - i\omega t\right) \cdot \exp\left(\gamma t\right)$$

We will consider only the effect of the electric field (directed along the X-axis) on the drifting electron in the quantum well located at $y \approx R$.

In the hydrodynamic approximation, functions $v\ (x,t)$ and $n\ (x,t)$ can be described by the motion equation

$$\frac{\partial v}{\partial t} + (v \cdot \nabla)v = \frac{e}{m}\left(E_{app} + F^{(1)}\right) - \frac{v}{\tau} - \frac{\nabla p}{m \cdot n} \tag{9}$$

and the equation of continuity

$$\frac{\partial n}{\partial t} + div(n\ v) = 0. \tag{10}$$

Here, e and m are the electron charge and the electron effective mass, respectively; E_{app} is the applied dc electric field, $F^{(1)}$ is electric field of the wave in a waveguide, τ^{-1} is the electron collision frequency. The last term on the right-hand side of Eq. (7) describes the effect of the pressure gradient (∇p) in a non-uniform gas of carriers.

Let us assume that $n(x,t) = n_0 + \delta n(x,t)$ and $v(x,t) = v_0 + \delta v(x,t)$, where n_0 and v_0 are the unperturbed concentration and velocity of the charge carriers. At $\tau^{-1} >> \omega$, the left side of Eq. (9) can be neglected, then

$$\delta v = \mu F_1 - D\frac{1}{n_0}\frac{\partial\ \delta n}{\partial x}, \tag{11}$$

where μ is mobility and D is diffusion coefficient of the charge carrier system in the quantum well.

We solve the system of Eqs. (9) and (11) for the boundary conditions $\delta n(0,t) = 0$ (that is, at the point of entry into the interaction space, the electron flow is uniform). In the linear approximation over $\delta n(x,t)$ and $\delta v(x,t)$, solutions take the form (in the region outside the boundary $x = 0$).

$$\delta n \approx \frac{n_0 k_c \mu E_0 C_1^{(1)} \exp\left(-k_c R\right)}{\delta^2 + \left(\gamma + D \cdot k_c^2\right)^2}\left\{-\delta \cos\left(k_c x - \omega t\right) - \left(Dk_c^2 + \gamma\right)\sin\left(k_c x - \omega t\right)\right\}\exp\left(\gamma t\right),$$

$$\tag{12}$$

$$\delta v \approx \frac{\mu E_0 C_1^{(1)} \exp\left(-k_c R\right)}{\delta^2 + \left(\gamma + Dk_c^2\right)}\left\{\left(Dk_c^2 \gamma + \delta^2\right)\cos\left(k_c x - \omega t\right) + Dk_c^2 \delta \sin\left(k_c x - \omega t\right)\right\}\exp\left(\gamma t\right).$$

$$\tag{13}$$

Here, $\delta = k_c v_0 - \omega$, $\delta << \omega$. The expressions (12)–(13) were obtained for condition $\gamma << Dk_c^2$, which is satisfied for typical semiconductors with diffusion coefficient $D > 1$ cm^2/s.

Amplification of electromagnetic waves is achieved due to the fact that drifting charge carriers transfer their energy to the radiation field. Represent the current of carriers, which interact resonantly with the wave, as the sum.

$J(x, \ t) = J_n(x, \ t) + J_v(x, \ t)$, where $J_n(x, \ t) = e \cdot \delta n(x, \ t) \cdot v_0 \cdot L_z$ is the current of the electron density wave, and $J_v(x, \ t) = e \cdot \delta v(x, \ t) \cdot n_0 \cdot L_z$ is the current due to velocity modulation of the electron stream, L_z is the length of the plate. (Usually, J_v is much smaller than J_n.) The electron current does work on the electromagnetic wave, and gives its energy to the electromagnetic wave. Let P denote the energy transferred by electrons to an electromagnetic wave per unit time and per unit area of the plate, and in addition per unit path of the electron flow averaged over the distance traveled. Using (12) and (13), we obtain:

$$P \approx - (1/2) \cdot e \cdot n_0 \cdot \left(C^{(1)} E_0 e^{-k_c R} \right)^2 \mu \cdot \theta, \qquad (14)$$

where the function

$$\theta = \frac{\delta \cdot (k \cdot v_0 + \delta)}{\delta^2 + \left(Dk_c^2 \right)^2}$$

characterizes the efficiency of energy transfer from the electron density waves to the electromagnetic wave. From θ it follows that if the electron velocity v_0 is strictly equal to ω/k_c (this corresponds to $\delta = 0$), the amplification of an electromagnetic wave is absent. Indeed, in the case of $\delta = 0$, electromagnetic wave energy, expended for electron acceleration, is equal to the energy that is returned to the wave by decelerating electrons. Thus, there is no amplification of an electromagnetic wave. In order for the wave amplification mode to occur, the electron velocity must be slightly higher than ω/k_c ($\delta > 0$).

Figure 5 shows θ as a function of δ, where $\delta = k_c v_0 - \omega$. The region of values of δ, in which the function θ takes positive values (and $P < 0$), corresponds to the amplification of the electromagnetic wave. The electromagnetic wave attenuation occurs for negative values of θ ($P > 0$). The maximum energy transfer from

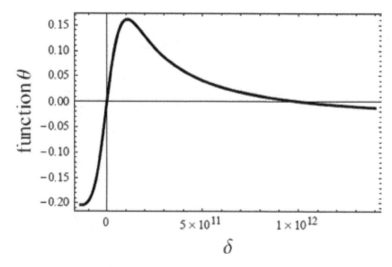

Figure 5.
θ as a function of $\delta = k_c v_0 - \omega$. The function θ characterizes the efficiency of energy transfer from electrons to the electromagnetic wave.

electrons to the wave (the maximum of the electromagnetic wave amplification) corresponds to the maximum of the function θ which is attained at

$$\delta_g \approx Dk_c^2\left(1 - \frac{Dk_c}{v_0}\right).$$

We carried out numerical estimates for the structure with p-Si conducting channel. The semiconductor p-Si has quite high values of the electrical breakdown field strength $(3 \cdot 10^5$ V/cm$)$, and high values of the saturation electron velocity $(1.5 \cdot 10^7$ cm/s$)$ in high electric fields and relatively low electron mobility at room temperature [10]. The crystal heterostructures $Si\ O_2/Si\ /Si\ O_2$ are mastered well enough by now [11].

For our estimates, we used the values of $\mu = 10^2\ cm^2 V^{-1}s^{-1}$ for room temperature mobility and $v_0 = 5 \cdot 10^6\ cm/s$ for the drift velocity in the applied field $E_{app} \approx 5 \cdot 10^4\ V/cm$. Then, at the frequency $\omega = 10^{12}\ s^{-1}$ and the wave number $k_c = \omega/v_0 = 2 \cdot 10^5\ cm^{-1}$ (which corresponds to the corrugation period $L_c = 0.3\ \mu$), we have $D \cdot k_c^2 \approx 1.2 \cdot 10^{11} s^{-1}$. The hole collision frequency is $\tau^{-1} = 10^{14}\ s^{-1}$, so condition $\tau^{-1} >> \omega$ is satisfied.

For amplification coefficient $\gamma = 10^8\ s^{-1}$ $(\gamma << D \cdot k_c^2)$, we have found the optimum mismatch $\delta_g \approx 10^{11}\ s^{-1}$, at which for the value of the energy transmitted from the electrons to the electromagnetic wave have its maximal in linear amplification regime.

The plate width $L_x = 2\pi/\chi \approx 0.87$ mm is chosen to be equal to half the natural wavelength with the frequency $\omega = 10^{12}\ s^{-1}$. With the above parameters, a maximum value of the field E_{max} reaches value of 10^4 V/cm at which the exponential growth of electromagnetic field comes to set end, and the process goes to a quasi-stationary state. For our estimates, we used $E_0 \approx 170$ V/cm, for which a linear regime of generation is certainly satisfied.

Negative energy flow of the electron density wave (represented by the expression (14)) is converted into a positive energy flow of the electromagnetic wave.

The change per unit time of the electromagnetic field energy of the open waveguide is determined by the sum $W_v + W_d$ of the contributions of the field energy outside the plate (in vacuum)

$$W_v \approx \frac{1}{4\pi} \cdot S \cdot \gamma \cdot e^{2\gamma t} \cdot E_0^2 \frac{c^2}{\omega^2 a}\frac{\varepsilon}{(\varepsilon - 1)},$$

and the field energy concentrated inside the plate

$$W_d \approx \frac{1}{4\pi} \cdot S \cdot \gamma \cdot e^{2\gamma t} \cdot E_0^2 \cdot \left[(\varepsilon + 1) \cdot \left(\frac{\cos a\chi Sin\, a\chi + a\chi}{\chi}\right)\right],$$

where $S = L_x L_z$ is the area of the plate.

For a structure with a plate of quartz and sapphire with the thickness $2a = 400\ \mu$, the wave vector takes the value of $s \approx a\omega^2(\varepsilon - 1)/(\varepsilon c^2) = 1.8 \cdot\ cm^{-1}$ in vacuum and the value of $\chi \approx (\varepsilon - 1)^{1/2}\omega/c = 72\ 3\ cm^{-1}$ in dielectric. In this case, the ratio of the wave energy in the dielectric to the wave energy in vacuum is $W_d/W_v \approx 10^{-3}$.

To start the electromagnetic oscillations, the gain γ must be larger than the loss factor

$$\gamma_{loss} = \omega \cdot tg \ \beta \cdot \frac{W_d}{W_v + W_d},$$

which characterizes the attenuation of waves due to dielectric losses. Here, $tg \ \beta$ is dielectric loss tangent, which depends on dielectric materials. For the structure of quartz and sapphire ($tg \ \beta \sim 10^{-4}$) with thickness $2 \cdot a = 400$ μ, and at frequency $\omega = 10^{12}$ s^{-1}, the value of the dielectric loss coefficient is $\gamma_l \approx 3 \cdot 10^6$ s^{-1}. Thus, the start-up condition $\gamma >> \gamma_{loss}$ is fulfilled with a sufficient margin.

The carrier density n_0, which can ensure the energy flow of the charge density wave for generation of terahertz radiation, is determined by the law of conservation: $P = W_v + W_d$. For the structures with the above parameters, we have $n_0 \approx 2 \cdot 10^{12}$ cm^{-2}. Then the generation power is 25 mW for plate area S = 1 cm^2.

The formation and amplification of the charge density wave are taking place in the system of the scattering carriers. Let us denote by Q the energy that drifting carriers lose in collisions per unit time:

$$Q = e n_0 \mu \cdot E_{app}^2 \cdot L_x \cdot L_z.$$

The efficiency of the generator can be estimated from the ratio of the useful energy P to the total energy $P + Q$ expended by the electrons both for amplification of electromagnetic waves and for collisions: $eff = |P|/(|P| + Q)$.

For the scheme with the above parameters, the generation efficiency is 1%, and the generation power is 25 mW at frequency of $\omega = 10^{12}$ s^{-1}.

Note that one of the possible ways to improve the output power is to increase the area of the plate. Moreover, the output power can be increased (according to expression (14)) by increasing the average electron concentration in the channel, as well as increasing the electron mobility. However, the mobility cannot be very high, since in the framework of the model described above, condition $\tau^{-1} >> \omega$ must be fulfilled (see conditions for Eq. (11) above). But the generation efficiency for the considered circuit does not depend on the plate area, electron concentration, and mobility.

4. Conclusion

Currently, there are many approaches to the problem of terahertz generator creation, and there are working devices. However, the problem is still not solved completely. Now efforts are focused not on the development of a single device with record-breaking parameters, but on devices that are suitable for wide application.

Here we consider the concept of a compact terahertz generator, for which neither low temperatures nor strong magnetic fields are required. So, it may have some advantages compared other terahertz generator.

We propose a terahertz generator scheme, which is an open waveguide with a thin dielectric plate, since in such a waveguide the ohmic losses of electromagnetic waves can be small. Indeed, in such waveguides, ohmic losses are present only in the dielectric plate, while the wave energy is concentrated mainly outside the plate.

The device plate consists of two dielectric layers with different values of permittivity and dielectric interface, periodically corrugated in the transverse direction of the waveguide. The transverse TE wave propagating in the waveguide causes polarization of the dielectric in the zone of the corrugation, and this polarization, in turn, induces an alternating electric field near the corrugated dielectric interface. An electron conducting channel is included in one of the dielectric layers of the plate close to the corrugated dielectric interface. Electrons drift in the conducting channel in the transverse direction of the waveguide (in the direction perpendicular to the direction of wave propagation). The drifting electrons interact with the inhomogeneous electric field which is induced near the dielectric interface by the TE wave of the waveguide. The corrugation period and the parameters of the electronic system are selected in such a way as to ensure the regime of the most effective interaction of drifting electrons with an electromagnetic wave. The gener-ation and amplification of terahertz waves occurs due to the efficient energy exchange between electrons and the electromagnetic wave of the waveguide [7]. It is important that in the considered scheme, the interaction of the electron flow with the electromagnetic wave is quite effective without slowing down the wave propa-gating in the waveguide.

One of the advantages of the proposed scheme is its low sensitivity to the spread of heterostructure parameters. Before fabrication the corrugated structure, it is possible to define corrugation parameters, under which the synchronism condition will be satisfied.

We have calculated the induced inhomogeneous electric field of the wave near the corrugated dielectric interface, and we have obtained the optimal values of the corrugation amplitude and period at which the amplitude of the non-uniform electric field takes its maximum.

In the hydrodynamic approximation, we have obtained analytical solutions of the equations taking into account the diffusion and drift of the charge carriers in the quantum well. The solutions describe density wave of charge carriers that interact with the first spatial harmonic wave of an inhomogeneous electric field near the corrugation. We have presented numerical estimates for a structure with a plate of quartz and sapphire and the silicon conductance channel. We have found that it is possible to generate electromagnetic waves at the frequency $\omega = 10^{12}$ s^{-1} with the output power of 25 mW and efficiency 1% in waveguides with plate area 1 cm^2 and the average electron concentration in the channel $2 \cdot 10^{12}$ cm^{-2}.

Author details

Ljudmila Shchurova[1*] and Vladimir Namiot[2]

1 Division of Solid State Physics, Lebedev Physical Institute, Russian Academy of Science, Moscow, Russia

2 Department of Microelectronics, Scobelthsyn Research Institute of Nuclear Physics, Moscow State University, Moscow, Russia

*Address all correspondence to: ljusia@gmail.com

References

[1] Hübers HW, Richter H, Wienold M. High-resolution terahertz spectroscopy with quantum-cascade lasers. Journal of Applied Physics. 2019;**125**:15140. DOI: 10.1063/1.5084105

[2] Son JH, Oh SJ, Cheon H. Potential clinical applications of terahertz radiation. Journal of Applied Physics. 2019;**125**:190901. DOI: 10.1063/ 1.508 0205

[3] Dhillon SS, Vitiello MS, Linfield EH, et al. The 2017 terahertz science and technology roadmap. Journal of Physics. 2017;**D50**:043001

[4] Li D, Nakajima M, Tani M, Yang J, Kitahara H, Hashida M, et al. Terahertz radiation from combined metallic slit arrays. Scientific Reports. 2019;**9**:6804. DOI: 10.1038/s41598-019-43072-2

[5] Namiot VA, Shchurova LY. On the generation of electromagnetic waves in the terahertz frequency range. Physics Letters. 2011;**A375**:2759-2766

[6] Namiot VA, Shchurova LY. Possibility of using "sliding" particles in accelerators and tunable source of coherent ultraviolet radiation. Physics Letters. 2012;**A376**:3445-3453

[7] Shchurova LY, Namiot VA. On generation of electromagnetic waves in the terahertz frequency range in a multilayered open dielectric waveguide. Physics Letters. 2013; **A377**:2440-2446

[8] Shchurova L, Namiot V, Sarkisyan D. A compact source of terahertz radiation based on interaction of electrons in a quantum well with an electromagnetic wave of a corrugated waveguide. Biophysics. 2015;**60**:647-655

[9] Sveshnikov AG. Mathematical modeling in electrodynamics. Computational Mathematics and Modeling. 1995;**6**:254-263

[10] Levinshtein M, Rumyantsev S, Shur M, editors. Handbook Series on Semiconductor Parameters. New Jersey: World Scientific; 1996. 32 p. DOI: 10.1142/9789812832078_0001

[11] Lu ZH, Grozea D. Crystalline Si/ SiO2 quantum wells. Applied Physics Letters. 2002;**80**:255. DOI: 10.1063/1.143 3166

5

Recent Advancement on the Excitonic and Biexcitonic Properties of Low-Dimensional Semiconductors

Anca Armășelu

Abstract

Knowing excitonic and biexcitonic properties of low-dimensional semiconductors systems is extremely important for the discovery of new physical effects and for the development of novel optoelectronics applications. This review work furnishes an interdisciplinary analysis of the fundamental features of excitons and biexcitons in two-dimensional semiconductor structures, one-dimensional semiconductor structures, and zero-dimensional (0D) semiconductor structures. There is a focus on spectral and dynamical properties of excitons and biexcitons in quantum dots (QDs). A study of the recent advances in the field is given, emphasizing the latest theoretical results and latest experimental methods for probing exciton and biexci-ton dynamics. This review presents an outlook on future applications of engineered multiexcitonic states including the photovoltaics, lasing, and the utilization of QDs in quantum technologies.

Keywords: excitonic states, biexcitonic states, multiexcitonic states, reduced dimensionality semiconductors, quantum wells, quantum wires, quantum dots, applications

1. Introduction

The scientific significance in the field of the physics of excitons comprises both basic research and applied research, this area of physics being one of the most actively studied subjects. The interest in the physics of excitons has raised actively over the past two decades, this interest being provoked by the unique properties of excitons that provide the development's context of optoelectronic and photovoltaic of various device applications such as electrically driven light emitters [1–3], photovoltaic solar cells [4–6], photodetectors [7, 8], and lasers [9–12].

An important concern for the researchers of this field is to obtain a decrease of the dimension of the macroscopic semiconductor systems to nanoscale, this thing leading not only to manufacturing and observing low-dimensional semiconductor structures (LDSs) but also to the emergence and development of new electronic and optical properties that are significantly different from bulk semiconductors properties. Because the essential distinction between low-dimensional semiconductor structures and bulk semiconductor structures can be explained using the terminology of improved excitonic effects that are determined by exciton localization result, it is crucial that, in these quantum-confined materials, the excitonic properties to be better understood and mastered to be able to use them in the development of the innovative and proficient optoelectronic devices utilizing these types of the materials.

Numerous researches have already shown that exploring the excitons' behavior in low-dimensional semiconductor systems can find new ways of controlling the fundamental exciton properties for light generation and light harvesting and finding novel materials for the next-generation high-efficiency excitonic light-harvesting tools at low cost [13–15].

Excitons in low-dimensional semiconductor structures have been widely investigated latterly. A low-dimensional semiconductor structure is a system which presents quantum confinement effects, the movement of electrons or other par-ticles (holes, excitons, etc.) being limited in one or more dimensions.

The promising area of excitonics represents the science and manufacturing of the excitons in disordered and low dimensionality semiconductors (organic semiconductors, hybrid perovskites, colloidal semiconductor nanoparticles) [16–23] and guarantees much quicker efficiency of harmonizing with fiber optics, realizing some novel stages to perform exciton-based computation at room temperatures [24].

It is known that the absorption by a semiconductor of a photon with energy equal to or greater than its bandgap stimulates an electron from the valence band into the conduction band, the vacancy left behind in the valence band being characterized as a hole which is a quasiparticle carrying positive charge. The Coulomb attraction type between these particles with electrical charges of opposite sign provides a quantum structure of electron–hole pair type which is electrical neutral, called exciton. Excitons have numerous characteristics similar to those of atomic hydrogen [25, 26]. Using this type of hydrogen atom model, in crystalline materials, two types of excitons can be discussed in the two limiting cases of a small dielectric constant when the exciton is tightly bound Frenkel-like (the electron and the hole are tightly bound; the Coulomb interaction is poorly screened) in contrast with large dielectric constant when the exciton is weakly bound Wannier-like (the Coulomb interaction is strongly screened by the valence electrons, the electron, and the hole being weakly bound) [27–32]. For a semiconductor exciton named Wannier exciton which has a radius greater than lattice spacing, the effective-mass approxi-mation can be used [33–40].

The entire gamut of low-dimensional semiconductor systems comprises quantum dots (QDs) or zero-dimensional (0D) systems if the excitons are dimensionally confined in all directions, quantum wires (QWRs) or one-dimensional systems (1D) if they are semiconductor nanocrystals in which the excitons are confined only in the diameter direction and quantum wells (QWs), or two-dimensional systems (2D) if the quantum confinement occurs in the thickness direction, while the particle motion is free in the other two directions [41–43].

It has been shown that in the quantum confinement conditions, the size and shape of semiconductor nanocrystals show an influence on the exciton fine structure, this being presented like the mode in which the energetic states of the exciton are divided by crystal field asymmetry consequences and low-dimensional semiconductor structures shape anisotropy [41–50].

Besides the hydrogen characteristics of the exciton, it is known that in QWs, QWRs, and QDs, there are hydrogen atom-like exciton pair-state populations or larger bound systems called biexcitons [25–27, 51–56].

Various researchers have shown that with the rise of the exciton binding energy value in low-dimensional semiconductor systems, the biexciton binding energy value with growth confinement is also raised [25, 57–67]. All of the papers in this field have shown that by improving the biexciton creation in reduced dimensional semiconductor structures, the quantum yield (QY) of photovoltaic cells has been enhanced [57, 68–71]. Also, biexcitons are important for quantum-information and computation areas due to their stunning benefit for the creation of coherent combination of quantum states, in this sense being used to find new platforms for the obtaining of future and scalable quantum-information applications such as some greater efficiency non-blinking single-photon sources of biexciton, entangled light sources, and laser based on biexciton states [72–75].

Multiple exciton generation (MEG) in low-dimensional semiconductors is the procedure by which multiple electron–hole pairs, or excitons, are created after the absorption of a single high-energy photon (larger than two times the bandgap energy) and is an encouraging research direction to maximize the solar energy conversion efficiencies in semiconductor solar cells at a possibly much diminished price [76–78]. Numerous studies have shown that the photo-physical properties of MEG are due to the character of inherent multiexciton interaction [79, 80].

This present chapter reviews the recent advancement in the understanding of the excitons' and biexcitons' behavior in LDSs, this fact being important for new experiments and optoelectronic devices.

The second section of this paper comprises three important parts that analyze the way in which the properties of excitons and biexcitons in two-dimensional structures, one-dimensional semiconductor structures, and zero-dimensional semiconductor structures are influenced by the nanometric dimensions case.

The final section recapitulates the fundamental and special issues that have been debated.

2. Excitonic and biexcitonic properties in low-dimensional semiconductors

This part of the chapter contains some crucial and novel concepts of excitonic and biexcitonic properties of semiconductor structures of low dimensionality (e.g., QWs, QWRs, QDs) which are relevant for the characterization of the active con-stituents in advanced tools.

2.1 Excitons and biexcitons in two-dimensional semiconductor structures

This section presents a subject of an enormous significance for the excitons and biexcitons effects in two-dimensional semiconductor structures. In 2005, Klingshirn [25] reported some essence results which emphasize the optical proper-ties of excitons in QWs, in coupled quantum wells (CQWs), and superlattices.

In the last years, in the area of excitons in LDSs, there has been much study which integrates experimental, theoretical, and technical features about the effective-mass theory of excitons and explains numerical procedures to compute the optical absorption comprising Coulomb interaction cases [81–84].

Xiao and coworkers [85, 86] emphasized the case of the excitons functioning in some layered two-dimensional (2D) semiconductors, presenting new different methods of the obtaining of some propitious materials structure (like molybdenum disulfide MoS_2) with perfect properties for the evolving of the operable optoelec-tronics and photonics such as light-emitting diodes (LEDs), lasers, optical modula-tors, and solar cells based on 2D materials. In Ref. [87] the study of the enhanced Coulomb interactions in WSe_2-$MoSe_2$-WSe_2 trilayer van der Waals (vdW) hetero-structures via neutral and charged interlayer excitons dynamics is mentioned.

In the situation of cryogenic temperatures, an increasing photoluminescence quantum yield in the conditions of the inclusion of a WSe_2 layer in the trilayer composition in contrast with the example of the bilayer heterostructures has been reported.

Owing to the fact that the class of 2D materials presents some distinctive features, which are highly dissimilar in comparison with those of their three-dimensional (3D) correspondents, it is used for the next-generation ultra-thin electronics [88]. In this context some researchers explained the role of the expansion of indirect excitons (an indirect exciton—IX—is a bound pair of an electron and hole in separated QW layers [89]), which is observed in vdW transition metal dichalcogenide (TMD) heterostructures at room temperature, this study helping for the progress of excitonic devices with energy-productive computation and ideal connection quality for optical communication cases [90, 91]. Various theoretical and experimental analyses have been developed for the improvement of the exci-tonic devices that use IXs propagation in different types of single QWs and coupled QWs [92, 93]. Fedichkin and his colleagues [94] studied a novel exciton transport model in a polar (Al, Ga)N/GaN QWs calculating the propagation lengths up to 12 μm at room temperature and up to 20 μm at 10 K.

In Ref. [95] a theoretical portrayal of the ground and excited states of the excitons for GaAs/AlGaAs and InGaAs/GaAs finite square QWs of different widths is presented which eases the elucidation of the experimental reflectance and photolu-minescence spectra of excitons in QWs.

Some works examined new different excitonic properties of the 2D organic–inorganic halide perovskite materials showing that this type of perovskite is very qualified to be used for the construction of the photonics devices [96–98] containing LEDs [99, 100], photodetectors [101], transistors, and lasing applications [102]. Wang et al. [103] provided a valuable research about the special characteristics of the long-lived exciton, trion, and biexciton cases in CdSe/CdTe colloidal QWs, proposing a novel model of light harvester with minimal energy losses.

2.2 Excitons and biexcitons in one-dimensional semiconductor structures

One-dimensional semiconductor structures have obtained a remarkable consideration within the last decade. 1D semiconductor nanostructures including wires, rods, belts, and tubes possess two dimensions smaller than 100 nm [104]. Among these types of 1D nanostructures, semiconductor QWRs have been investigated thoroughly for a broad range of materials. This type of 1D nanostructures is used for an essential study due to their exclusive constitutional and physical properties comparative with their bulk correspondents. Crottini et al. [105] communicated the 1D biexcitons behavior in high-quality disorder-free semiconductor QWRs, evalu-ating the biexciton binding energy value at 1.2 meV.

Sitt and his coworkers [106] reviewed the excitonic comportment of a diversity of heterostructured nanorods (NRs) which are used for a series of applications comprising solid-state lighting, lasers, multicolor emission, bio-labeling, photon-detecting devices, and solar cells. In the same work [106], some multiexciton effects are shown, and the dynamics of charge carriers is presented in core/shell NRs with potential applications in the optical gain field and in the light-harvesting section.

For the case of the single crystalline silicon nanowires (SiNWs), which is a key structure for nanoscale tools including field-effect transistors, logic circuits, sensors, lasers, Yang [107] described some excitonic effects and the case of the optical absorption spectra using the Bethe-Salpeter equation.

In Ref. [108] the physical properties of elongated inorganic particles are reported in the case of the nanoparticle shape modification from spherical to rod-like with the help of the exciton storage process.

2.3 Excitons and biexcitons in zero-dimensional semiconductor structures

Zero-dimensional semiconductor structures have captivated a notable interest owing to the fact that the motion is confined in all three directions, the size of a QDs being smaller than or comparable to the bulk exciton Bohr radius [109–112]. In this part of the chapter, some recent progresses in the topic which deals with the excitonic and biexcitonic effects for QDs applications case are emphasized, including computing and communication field, light-emitting devices, solar cells area, and biological domain [113, 114].

Pokutnyi [115–120] realized the foremost theoretical analyses that accurately describe different absorption mechanisms in such nanosystems, discussing many issues related to the complicated interrelationship between the morphology of the zero-dimensional semiconductor structures and their electronics and their optical properties and which help to the progress of novel proficient optoelectronic devices.

Plumhof and his colleagues [121] proved that QDs with an adequately small excitonic fine structure splitting (FSS) can be utilized as some valuable deterministic sources of polarization-entangled photon pairs to improve the building blocks' quality for quantum communication technology.

Golasa et al. [122] presented some new statistical properties of neutral excitons, biexcitons, and trions for the case of QDs which are created in the InAs/GaAs wetting layer (WL), confirming that the WLQDs structure is a useful model to be applied in the area of quantum-information processing applications.

Singh and his team of researchers [123] found a new multipulse time-resolved fluorescence experiment for the CdSe/CdS core/shell QDs case, this work being a crucial spectroscopic procedure which can separate and measure the recombination times of multiexcited state for the proposed sample.

In Ref. [124] a comprehensive review is furnished about the appropriately engineered core/graded-shell QDs revealing advantageous optical properties and unique photoluminescence assets of QDs for liquid crystal displays backlighting technologies and organic light-emitting diode tools. In different papers which have to do with the quantum dot-based-light-emitting diodes (QD-LEDs) results [124–127], it is mentioned that for the improvement of the QD-LED performance, two processes must be diminished: trapping of carriers at surface defects and Auger recombination of excitons.

Important studies reveal many novel and interesting experimental and theoretical results on LDSs exhibiting high quantum yield as a result of MEG occurrence with the aim of the improvement of the solar devices field. Considering that Shockley and Queisser determined a basic threshold value for the efficiency of a traditional p-n solar cell of 30%, these essential results prove that there is a possibility to exceed the Shockley-Quiesser threshold employing quantum effects for a recently developed low-cost third-generation solar cell [77, 128–130].

3. Conclusions

In recent decades, low-dimensional semiconductor structures have become one of the most dynamic research areas in nanoscience, the excitons showing some notably novel attributes due to confinement consequence case. In this chapter a review of some modern experimental and theoretical discoveries on excitonic and biexcitonic effects in low-dimensional semiconductors is presented. The paper furnishes an outstandingly multipurpose excitonic aspect of the optoelectronic applications field, including photodetectors and opto-valleytronic tools, computing and communication domain, and light-emitting devices.

Author details

Anca Armășelu
Faculty of Electrical Engineering and Computer Science, Department of Electrical
Engineering and Applied Physics, Transilvania University of Brasov, Brasov,
Romania

*Address all correspondence to: ancas@unitbv.ro

References

[1] Mueller T, Malic E. Exciton physics and device application of two-dimensional transition metal dichalcogenide semiconductors. npj 2D Materials and Applications. 2018;2(29):1-12. DOI: 10.1038/s41699-018-0074-2

[2] Paur M, Molina-Mendoza JA, Bratschitsch R, Watanabe K, Taniguchi T, Mueller T. Electroluminescence from multi-particle exciton complexes in transition metal dichalcogenide semiconductors. Nature Communications. 2019;10(1):1709. DOI: 10.1038/s41467-019-09781-y

[3] Ko YK, Kim JH, Jin LH, Ko SM, Kwon BJ, Kim J, et al. Electrically driven quantum dot/wire/well hybrid light-emitting diodes. Advanced Materials. 2011;23(45):5364-5369. DOI: 10.1002/adma.201102534

[4] Beard MC, Luther JM, Semonin OE, Nozik AJ. Third generation Photovoltaics based on multiple Exciton generation in quantum confined semiconductors. Accounts of Chemical Research. 2013;46(6): 1252-1260. DOI: 10.1021/ar3001958

[5] Głowienka D, Szmytkowski J. Influence of excitons interaction with charge carriers on photovoltaic parameters in organic solar cells. Chemical Physics. 2018;503(31):31-38. DOI: 10.1016/j.chemphys. 2018.02.004

[6] Arkan F, Izadyar M. The role of solvent and structure in the porphyrin-based hybrid solar cells. Solar Energy. 2017;146(368):368-378. DOI: 10.1016/j. solener.2017.03.006

[7] Gong Y, Liu Q, Gong M, Wang T, Zeng G, Chan WL, et al. Photodetectors: High-performance Photodetectors based on effective exciton dissociation in protein-adsorbed multiwalled carbon nanotube Nanohybrids. Advanced Optical Materials. 2017;5(1). DOI: 10.1002/adom.201770002

[8] Tsai H, Asadpour R, Blancon JC, Stoumpos CC, Even J, Ajayan PM, et al. Design principles for electronic charge transport in solution-processed vertically stacked 2 D perovskite quantum wells. Nature Communications. 2018;9:1-9. DOI: 10.1038/s41467

[9] Jahan KL, Boda A, Shankar IV, Raju NC, Chatterjee A. Magnetic field effect on the energy levels of an exciton in a GaAs quantum dot: Application for exciton lasers. Scientific Reports. 2018;8(1):1-13. DOI: 10.1038/s41598-018-23348-9

[10] Fraser MD, Höfling S, Yamamoto Y. Physics and applications of exciton-polariton lasers. Nature Materials. 2016;15(10):1049-1052. DOI: 10.1038/nmat4762

[11] Höfling S, Schneider C, Rahimi- Iman A, Kim NY, Amthor M, Fischer J, et al. Electrically driven Exciton-Polariton lasers. In: Conference on Lasers and Electro-Optics (CLEO 2014)—Laser Science to Photonic Applications; 8-13 June 2014; San Jose, CA, USA. 2014. pp. 1-2. Electronic ISBN: 978-1-55752-999-2. Print ISSN: 2160-8989

[12] Pokutnyi SI. Optical nanolaser on the heavy hole transition in semiconductor nanocrystals: Theory. Physics Letters A. 2005;342:347-350

[13] Guzelturk B, Martinez PLH, Zhang Q, Xiong Q, Sun H, Sun XW, et al. Excitonics of semiconductor quantum dots and wires for lighting and displays. Laser & Photonics Reviews. 2014;8(1):73-93. DOI: 10.1002/lpor.20 1300024

[14] Frazer L, Gallaher JK, Schmidt TW. Optimizing the efficiency of solar

photon Upconversion. ACS Energy Letters. 2017;**2**(6):1346-1354. DOI: 10.1021/acsenergylett.7b00237

[15] Pokutnyi SI. Exciton states in semiconductor nanocrystals: Theory (review). Physics Express. 2011;**1**: 158-164

[16] La Société Française d'Optique. 2018. Available from: https://www.sfoptique.org/pages/ecoles-thematiques/excitonics-thematic-school

[17] Sum TC, Mathews N, editors. Halide Perovskites: Photovoltaics, Light Emitting Devices, and beyond. 1st ed. Weinheim: Wiley-VCH; 2019. p. 312. Print ISBN: 978-3-527-34111-5

[18] Efremov NA, Pokutnyi SI. The energy spectrum of the exciton in a small semiconductor particles. Soviet Physics - Solid State. 1990;**32**(6): 955-960

[19] Pokutnyi SI. Exciton states in semiconductor nanostructures. Semiconductors. 2005;**39**(9):1066-1070

[20] Pokutnyi SI, Ovchinnikov OV, Smirnov MS. Sensitization of photoprocesses in colloidal Ag_2S quantum dots by dye molecules. Journal of Nanophotonics. 2016;**10**:033505-1-033505-12

[21] Pokutnyi SI, Ovchinnikov OV. Relationship between structural and optical properties in colloidal CdxZn1-xS quantum dots in gelatin. Journal of Nanophotonics. 2016;**10**:033507-1-033507-13

[22] Pokutnyi SI, Ovchinnikov OV. Absorption of light by colloidal semiconducor quantum dots. Journal of Nanophotonics. 2016;**10**:033506-1-033506-9

[23] Pokutnyi SI. Optical absorption by colloid quantum dots CdSe in the

dielectric matrix. Low Temperature Physics. 2017;**43**(12):1797-1799

[24] Borghino D. New advances in excitonics promise faster computers. 2009. Available from: https://newatlas.com/excitonics-faster-computers-solar-panels/13010

[25] Klingshirn C. Semiconductor Optics. 2nd ed. Berlin: Springer Verlag; 2005. p. 797. ISBN: 3-540-21328-7

[26] Koch SW, Kira M, Khitrova G, Gibbs HM. Semiconductor excitons in new light. Nature Materials. 2006;**5** (7):523-531. DOI: 10.1038/nmat1658

[27] Yu YP, Cardona M. Fundamentals of Semiconductors. Physics and Materials Properties. 4th ed. Vol. 778p. Berlin: Springer Verlag; 2010. DOI: 10.1007/978-3-642-00710-1

[28] Neutzner S, Thorin F, Corteccihia D. Exciton-polaron spectral structures in two-dimensional hybrid lead-halide perovskites. Physical Review Materials. 2018;**2**(6):064605. DOI: 10.1103/physrevmaterials.2.064605

[29] Pokutnyi SI. Two-dimensional Wannier-Mott exciton in a uniform electric field. Physics of the Solid State. 2001;**43**(5):923-926

[30] Pokutnyi SI. Exciton formed from spatially separated electrons and holes in dielectric quantum dots. Journal of Advances in Chemistry. 2015;**11**: 3848-3852

[31] Pokutnyi SI, Kulchin YN, Dzyuba VP. Exciton spectroscopy of spatially separated electrons and holes in the dielectric quantum dots. Crystals. 2018;**8**(4):148-164

[32] Pokutnyi SI, Kulchin YN, Amosov AV, Dzyuba VP. Optical absorption by a nanosystem with

dielectric quantum dots. Proceedings of SPIE. 2019;**11024**:1102404-1-1102404-6

[33] Pokutnyi SI. Exciton states in semiconductor quantum dots in the framework of the modified effective mass method. Semiconductors. 2007; **41**(11):1323-1328

[34] Pokutnyi SI. Spectrum of exciton in quasi-zero-dimensional systems: Theory. Physics Letters A. 1995;**203**: 388-392

[35] Pokutnyi SI. The spectrum of an exciton in quasi-zero-dimensional semiconductor structures. Semiconductors. 1996;**30**(11):1015-1018

[36] Pokutnyi SI. Excitons in quasi-zero-dimensional structures. Physics of the Solid State. 1996;**38**(2):281-285

[37] Pokutnyi SI. Exciton states in quasi-zero-dimensional semiconductor nanosystems: Theory (Review). Physics Express. 2012;**2**:20-26

[38] Pokutnyi SI. Exciton states in semiconductor quasi-zero-dimensional nanosystems. Semiconductors. 2012;**46**(2):165-170

[39] Pokutnyi SI. On an exciton with a spatially separated electron and hole in quasi-zero-dimensional semiconductor nanosystems. Semiconductors. 2013;**47**(6):791-798

[40] Pokutnyi SI, Kulchin YN, Dzyuba VP. Optical absorption by excitons of quasi-zero-dimensional dielectric quantum dots. Proceedings of SPIE. 2016;**10176**:1017603-1-1017603-7

[41] Rabouw FT, Donega CM. Excited-state dynamics in colloidal semiconductornanocrystals. In: Credi A, editor. Photoactive Semiconductor Nanocrystal Quantum Dots. Fundamentals Semiconductor Nanocrystal Quantum Dots.

Fundamentals and Applications. Cham: Springer; 2016. p. 179. DOI: 10.1007/ 978-3-319-51192-4

[42] Das P, Ganguly S, Banerjee S, Das NS. Graphene based emergent nano-lights: A short review on the synthesis, properties and application. Research on Chemical Intermediates. 2019;**45**(7): 3823-3853. DOI: 10.1007/s11164-019-03823-2

[43] Pokutnyi SI, Kulchin YN. Special section guest editorial: Optics, spectroscopy and Nanophotonics of quantum dots. Journal of Nano photonics. 2016;**10**(3):033501-1- 0335 01-8

[44] Siebers B. Spectroscopy of Excitons in CdSe/CdS Colloidal Nanocrystals [Dissertation]. Dortmund: Faculty of Physics of TU Dortmund University, Germany; 2015

[45] Pokutnyi SI. Spectroscopy of quasiatomic nanostructures. Journal of Optical Technology. 2015;**82**:280-285

[46] Pokutnyi SI. Size quantization of electron-hole pair in quasi-zero-dimensional semiconductor structures. Semiconductors. 1991;**25**(4):381-385

[47] Pokutnyi SI. Theory of size quantization of exciton in quasi-zero-dimensional semiconductor structures. Physica Status Solidi B. 1992;**173**(2): 607-613

[48] Pokutnyi SI. Spectrum of a size-quantized exciton in quasi-zero-dimensional structures. Physics of the Solid State. 1992;**34**(8):1278-1281

[49] Pokutnyi SI. Size quantization of excitons in quasi-zero-dimensional structures. Physics Letters A. 1992;**168**: 433-436

[50] Pokutnyi SI. Stark effect in semiconductor quantum dots. Journal

of Applied Physics. 2004;**96**:100015-100020

[51] Pokutnyi SI. The biexciton with a spatially separated electrons and holes in quasi-zero-dimensional semiconductor nanosystems. Semiconductors. 2013;**47**(12):1626-1635

[52] Pokutnyi SI, Kulchin YN, Dzyuba VP. Biexciton in nanoheterostructures of dielectric quantum dots.

Journal of Nanophotonics. 2016;**10**:036008-1-036008-8

[53] Pokutnyi SI. Biexciton in quantum dots of cadmium sulfide in a dielectric matrix. Technical Physics. 2016;**61**(11):1737-1739

[54] Pokutnyi SI. Biexciton in nanoheterostructures of germanium quantum dots. Optical Engineering. 2017;**56**(6):067104-1-067104-5

[55] Pokutnyi SI, Kulchin YN, Dzyuba VP. Biexciton states in nanoheterostructures of dielectric quantum dots. Journal of Physics Conference Series. 2018;**1092**(1):12029

[56] Pokutnyi SI. Size-quantized exciton in quasi-zero-dimensional structures. Physics of the Solid State. 1996;**38**(9):1463-1465

[57] Kershaw SV, Rogach AL. Carrier multiplication mechanism and competing processes in colloidal semiconductor nanostructures. Materials. 2017;**10**(9):1095. DOI: 10.3390/ma10091095

[58] Pokutnyi SI. The binding energy of the exciton in semiconductor quantum dots. Semiconductors. 2010;**44**:507-514

[59] Pokutnyi SI. Binding energy of the exciton with a spatially separated electron and hole in quasi-zero-dimensional semiconductor

nanosystems. Technical Physics Letters. 2013;**39**:233-235

[60] Pokutnyi SI. Binding energy of excitons formed from spatially separated electrons and holes in insulating quantum dots. Semiconductors. 2015;**49**(10):1311-1315

[61] Pokutnyi SI. Binding energy of a quasi-molecule in nanoheterostructures. Journal of Optical Technology. 2016;**83**(8):459-462

[62] Pokutnyi SI, Salejda W. Excitonic quasimolecules consisting of two semiconductor quantum dots: A theory. Ukrainian Journal of Physical Optics. 2016;**17**(3):91-97

[63] Pokutnyi SI, Salejda W. Excitonic quasimolecules in nanosystems containing quantum dots. Journal of Advances in Chemistry. 2015;**12**(2): 4018-4021

[64] Pokutnyi SI, Salejda W. Excitonic quasimolecules containing of two semiconductor quantum dots. Optica Applicata. 2016;**56**:629-637

[65] Pokutnyi SI. Excitonic quasimolecules containing of two quantum dots. Journal of Advanced Physics. 2016;**11**(9):4024-4028

[66] Pokutnyi SI, Gorbyk PP. New superatoms in alkali-metal atoms. Journal of Nanostructure in Chemistry. 2015;**5**(1):35-38

[67] Pokutnyi SI. Excitons based on spatially separated electrons and holes in Ge/Si heterostructures with germanium quantum dots. Low Temperature Physics. 2016;**42**(12): 1151-1154

[68] Ma X, Diroll BT, Cho W, Fedin I, Schaller RD, Talapin DV, et al. Size-dependent quantum yields and carrier dynamics of quasi-two-dimensional Core/Shell Nanoplatelets. ACS Nano.

2017;**11** (9) : 9119-9127. DOI: 10.1021/acsnano.7603943

[69] Pokutnyi SI, Kulchin YN, Dzyuba VP. Excitonic quasimolecules in quasi-zero-dimensional nanogeterostructures: Theory. Pacific Science Review A. 2016;**17**:11-13

[70] Pokutnyi SI. Excitonic quasimolecules formed by spatially separated electrons and holes in a Ge/Si heterostructure with germanium quantum dots. Journal of Applied Spectroscopy. 2017;**84**(2):268-272

[71] Pokutnyi SI. Exciton quasimolecules formed from spatially separated electrons and holes in nanostructures with quantum dots of germanium. Molecular Crystals and Liquid Crystals. 2018;**674**(1):92-97

[72] Chen J, Zhang Q, Shi J, Zhang S, Du W, Mi Y, et al. Room temperature continuous-wave excited biexciton emission in perovskite nanoplatelets via plasmonic nonlinear Fano resonance. Communications on Physics. 2019;**2**:80. DOI: 10.1038/s42005-019-0178-9

[73] Salter CL, Stevenson RM, Farres I, Nicoll CA, Ritchie DA, Shields AJ. Entangled-photon pair emission from a light-emitting diode. Journal of Physics: Conference Series. 2011;**286**(1):012022. DOI: 10.1088/1742-6596/286/012022

[74] Pokutnyi SI. Excitonic quasimolecules in nanosystems of semiconductor and dielectric quantum dots. Modern Chemistry & Applications. 2016;**4**(4):188-194

[75] Pokutnyi SI. Excitonic quasimolecules in nanosystems of quantum dots. Optical Engineering. 2017;**56**(9):091603-1-091603-7

[76] Siemons N, Serafini A. Multiple Exciton generation in nanostructures for advanced photovoltaic cells. Journal of Nanotechnology. 2017;**2018**(8):7285483. DOI: 10.1155/2018/7285483

[77] Beard CM, Ellingson RJ. Multiple exciton generation in semiconductor nanocrystals: Toward efficient solar energy conversion. Laser & Photonics Reviews. 2008;**2**(5):377-399. DOI: 10.1002/lpor.200810013

[78] Choi Y, Sim S, Lim SS, Lee YH, Choi H. Ultrafast biexciton spectroscopy in semiconductor quantum dots: Evidence for early emergence of multiple-exciton generation. Scientific Reports. 2013;**3**:3206. DOI: 10.1038/srep03206

[79] Smith C, Binks D. Multiple Exciton generation in colloidal Nanocrystals. Nanomaterials. 2014;**4**(1):19-45. DOI: 10.3390/nano4010019

[80] Ikezawa M, Nair SV, Ren HW, Masumoto Y, Ruda H. Biexciton binding energy in parabolic GaAs quantum dots. Physical Review B: Condensed Matter. 2006;**73**(12):12e53 21. DOI: 10.1103/PhysRevB.73.125321

[81] Viña I. Magneto-Excitons in GaAs/GaAlAs quantum Wells. In: Fasol G, Fasolino A, Lugli P, editors. Spectroscopy of Semiconductor Microstructures. 1st ed. New York: Springer US; 1989. p. 667. DOI: 10.1007/978-1-4757-6565-6

[82] Glutsch S. Excitons in Low-Dimensional Semiconductors. Theory Numerical Methods Applications. 1st ed. Berlin: Springer-Verlag; 2004. p. 298. DOI: 10.1007/978-3-662-07150-2

[83] Böer KW, Pohl UW. Excitons. In: Böer KW, Pohl UW, editors. Semiconductor Physics. 1st ed. Cham: Springer; 2017. DOI: 10.1007/978-3-319-06540-3_14-2

[84] Pokutnyi SI. Spectrum of a quantum - well electron - hole pair in

semiconductor nanocrystals. Semic-onductors. 1996;**30**(7):694-695

[85] Xiao J, Zhao M, Wang Y, Zhang X. Excitons in atomically thin 2D semiconductors and their applicat-ions. Nano. 2017;**6**(6):1309- 1328. DOI: 10.1515/nanoph-2016-0160

[86] Amani M, Lien DH, Kiriya D, Xiao J, Azcati A, Noh J, et al. Near-unity photoluminescence quantum yield in MoS_2. Science. 2015;**350** (6264):1065- 1068. DOI: 10.1126/ science.aad2114

[87] Choi C, Huang J, Cheng HC, Kim H, Vinod AK, Bae SH, et al. Enhanced interlayer neutral excitons and trions in trilayer van der Waals heter-ostructures. npj 2D Materials and Applications. 2018;**2**(1):30. DOI: 10.1038/s41699-018-0075-1

[88] Velický M, Toth PS. From two-dimensional materials to their heterostructures: An electrochemist's perspective. Applied Materials Today. 2017;**8**:68-103. DOI: 10.1016/ j. apmt.2017.05.003

[89] Lozovik YE, Yudson VI. A new mechanism for superconductivity: Pairing between spatially separated electrons and holes. Soviet Physics - JETP. 1976;**44**(2):738-753. Available from: http://www.jetp.ac.ru/cgi-bin/ dn/e_044_02_0389.pdf

[90] Calman EV, Fogler MM, Butov LV, Hu S, Mishchenko A, Geim AK. Indirect excitons in van der Waals heterostructures at room temperature. Nature Communica-tions. 2017;**9**(1):1985. DOI: 10.1038/ s41467-018-04293-7

[91] Dzyuba VP, Pokutnyi SI, Amosov AV, Kulchin YN. Indirect Excitons and polarization of dielectric nanoparticles. The Journal of Physical Chemistry C. 2019;**123** (42):26031-26035

[92] Leonard J. Exciton Transport Phenomena in GaAs Coupled Quantum Wells (Springer Theses). 1st ed. Cham: Springer International Publishing AG; 2018. p. 59. DOI: 10.1007/978-3-319-69733-8

[93] Butov LV. Excitonic devices. Superlattices and Microstructures. 2017;**108**:2-26. DOI: 10.1016/j. spmi. 2016.12.035

[94] Fedichkin F, Guillet T, Valvin P, Jouault B, Brimont C, Bretagnon T, et al. Room-temperature transport of indirect Excitons in (Al, Ga)N/GaN quantum Wells. Physical Review Applied. 2016;**6**:014011. DOI: 10.1103/ PhysRevApplied.6.014011

[95] Belov PA. Energy spectrum of excitons in square quantum wells. Physica E: Low-dimensional Systems and Nanostructures. 2019;**112**:96-108. DOI: 10.1016/j.physe.2019.04.008

[96] Mauck CM, Tisdale WA. Excitons in 2D organic-inorganic halide Perovskites. Trends in Chemistry. 2019;**1**(4):380-393. DOI: 10.1016/j.trechm.2019.04.003

[97] Misra RV, Cohen BE, Iagher L, Etgar L. Low-dimensional organic-inorganic halide Perovskite: Structure, properties an applications. ChemSus Chem. 2017;**10**:3712-3721. DOI: 10.1002/cssc.20170126

[98] Zhang R, Fan JF, Zhang X, Yu H, Zhang H, Mai Y, et al. Nonlinear optical response of organic-inorganic halide Perovskites. ACS Photonics. 2016;**3**(3):371-377. DOI: 10.1021/ acsphotonics.5b00563

[99] Zou W, Li R, Zhang S, Liu Y, Wang N, Cao Y, et al. Minimising efficiency roll-off in high-brightness perovskite light - emitting diodes. Nature Comm-unications. 2018;**9**(1):608. DOI: 10.10 38/s41467-018-03049-7

[100] Zhang L, Yang X, Jiang Q , Wang P, Yin Z, Zhang X, et al. Ultra-bright and highly efficient inorganic based perovskites light - emitting diodes. Nature Communications. 2017;**8**:15640. DOI: 10.1038/ncom-ms15640

[101] Li Y, Shi ZF, Li XJ, Shan CX. Photodetectors based on inorganic halide perovskites: Materials and devices. Chinese Physics B. 2019;**28**:017803. DOI: 10.1088/1674-1056/28/1/017803

[102] Stylianakis MM, Maksudov T, Panagiotopoulos A, Kakavelakis G, Petridis K. Inorganic and hybrid Perovskite based laser devices: A review. Materials. 2019;**12**(6):859. DOI: 10.3390/ma12060859

[103] Wang JH, Liang GJ, Wu KF. Long-lived single Excitons, Trions, and Biexcitons in CdSe/CdTe type-II colloidal quantum Wells. Chinese Journal of Chemical Physics. 2017;**30**(6):649-656. DOI: 10.1063/1674-0068/30/cjcp1711206

[104] Xia Y, Yang P, Sun Y, Wu Y, Mayers B, Gates B, et al. One-dimensional nanostructures: Synthesis characterization, and applications. Advanced Materials. 2003;**15**(5):353- 389. DOI: 10.1002/adma.200390087

[105] Crottini A, Staehli JL, Deveaud B, Wang XL, Ogura M. One-dimensional biexcitons in a single quantum wire. Solid State Communications. 2002;**121**(8):401-405. DOI: 10.1016/S0038-1098(01) 00510-5

[106] Sitt A, Hadar J, Banin U. Band-gap engineering optoelectronic properties and applications of colloidal heter-ostructured semiconductor nanorods. Nano Today. 2013;**8**(5):494- 513. DOI: 10.1016/j.nantod.2013.08.002

[107] Yang L. Excited-state properties of thin silicon nanowires. In: Andreoni W, Yip S, editors. Hand-book of Materials Modeling.

Cham: Springer; 2018. DOI: 10.1007/978-3-319-50257-1_37-1

[108] Krahne R, Morello G, Figuerola A, George C, Deka S, Manna L. Physical properties of elongated inorganic nanoparticles. Physics Reports. 2011;**501**(3-5):75-221. DOI: 10.1016/j.physrep.2011.01.001

[109] El-Saba M. Transport of Information-Carriers in Semiconduc-tors and Nanodevices. 1st ed. Hershey: IGI Global; 2017. p. 696. DOI: 10.4018/978-1-5225-2312-3

[110] Brkić S. Applicability of quantum dots in biomedical science. In: Djezzar B, editor. Ionizing Radiation Effects and Applications. 1st ed. London: IntechOpen; 2017. DOI: 105772/intechopen.68295

[111] Kulakovskii VD, Bacher G, Weigand R, Kümell T, Forchel A. Fine structure of Biexciton emission in symmetric and asymmetric CdSe/ZnSe single quantum dots. Physical Review Letters. 1999;**82**(8):1780-1783. DOI: 10.1103/PhysRevLett82.1780

[112] Pokutnyi SI. Exciton states formed by spatially separated electron and hole in semiconductor quantum dots. Technical Physics. 2015;**60**:1615-1618

[113] Pokutnyi SI. Quantum-chemical analysis of system consisting of two CdS quantum dots. Theoretical and Experimental Chemistry. 2016;**52**(1): 27-32

[114] Pokutnyi SI. Spectroscopy of excitons in heterostructures with quantum dots. Journal of Applied Spectroscopy. 2017;**84**(4):603-610

[115] Pokutnyi SI. Exciton states and direct interband light absorption in the ensemble of toroidal quantum dots. Journal of Nanophotonics. 2017;**11**(4): 046004-1-046004-11

[116] Pokutnyi SI. Strongly absorbing light nanostructures containing metal quantum dots. Journal of Nanophotonics. 2018;**12**(1):012506-1-012506-6

[117] Pokutnyi SI. Exciton states and optical absorption in core/shell/shell spherical quantum dot. Chemical Physics. 2018;**506**:26-30

[118] Pokutnyi SI. Optical spectroscopy of excitons with spatially separated electrons and holes in nanosystems containing dielectric quantum dots. Journal of Nanophotonics. 2018;**12**(2): 026013-1-026013-16

[119] Pokutnyi SI. Exciton spectroscopy with spatially separated electron and hole in Ge/Si heterostructure with germanium quantum dots. Low Temperature Physics. 2018;**44**(8): 819-823

[120] Pokutnyi SI. New quasi-atomic nanostructures containing exciton quasimolecules and exciton quasicrystals: theory. Surface. 2019; **11**(26):472-483. DOI: 10.1540/ Surface.2019.11.472

[121] Plumhof JD, Trotta R, Rastelli A, Schmidt OG. Experimental methods of post-growth tuning of the excitonic fine structure splitting in semiconductor quantum dots. Nanoscale Research Letters. 2012;**7**:336. DOI: 10.1186/1556-276X-7-336

[122] Gołasa K, Molas M, Goryca M, Kazimierczuk T, Smoleński T, Koperski M, et al. Properties of Excitons in quantum dots with a weak confinement. Acta Physica Polonica A. 2013;**124**(5):781. DOI: 10.12693/APhysPolA.124.781

[123] Singh G, Guericke MA, Song Q, Jones M. A multiple time-resolved fluorescence method for probing second-order recombination dynamics in colloidal quantum

dots. The Journal of Physical Chemistry C. 2014;**118**(26):14692- 14702. DOI: 10.1021/jp5043766

[124] Todescato F, Fortunati I, Minoto A, Signorini R, Jaseniak JJ, Bozio R. Engineering of semiconductor Nanocrystals for light emitting applications. Materials. 2016;**9**(8):672. DOI: 10.3390/ ma9080672

[125] Armășelu A. Recent developments in applications of quantum-dot based light-emitting diodes. In: Ghamsari MS, editor. Quantum-Dot Based Light-Emitting Diodes. 1st ed. London: IntechOpen; 2017. DOI: 10.5772/ intechopen.69177

[126] Bae WK, Park YS, Lim J, Lee D. Controlling the influence of auger recombination on the performance of quantum-dot light-emitting diodes. Nature Communications. 2013;**4**:2661. DOI: 10.1038/ncomms3661

[127] Zhou W, Coleman JJ. Semiconductor quantum dots. Current Opinion in Solid State and Materials Science. 2016;**20**(6):352-360. DOI: 10.1016/j.cossms.2016.06.006

[128] Goodwin H, Jellicoe TC, Davis NJLK, Böhm ML. Multiple exciton generation in quantum dot-based solar cells. Nano. 2017;**7**(1): 111-126. DOI: 10.1515/nanoph-2017-0034

[129] Beard MC. Multiple Exciton generation in semiconductor quantum dots. Journal of Physical Chemistry Letters. 2011;**2**(11):1282-1288. DOI: 10.1021/jz200166y

[130] Yan Y, Crisp RW, Gu J, Chernomordik BD, Pach GF, Marshall AR, et al. Multiple exciton generation for photoelectrochemical hydrogen evolution reaction with quantum yields exceeding 100%. Nature Energy. 2017;**2**:17052. DOI: 10.1038/nenergy.2017.52

6

Applications of Graphene Modified by Self-Assembled Monolayers

Gulsum Ersu, Yenal Gokpek, Mustafa Can, Ceylan Zafer and Serafettin Demic

Abstract

Self-assembled monolayers (SAMs) are well-oriented molecular structures that are formed by the adsorption of an active site of a surfactant onto a substrate's surface. Aromatic SAMs were used to modify anode/hole transport layer interface in order to achieve preferable barrier alignment and charge carrier injection from anode to an organic-based thin film material. Other functions of SAMs include current blocking layers or moisture penetration blocking layers, dipolar surface layers for enhanced charge injection, and modification of work function of a material such as graphene acting as a spacer to physically separate and electrically decouple it from the substrate. Additionally, SAM modification of graphene leads to its electronic passivation at layers' edges, elimination of defects, and enhanced adhesion and stability. The surface modification with molecules capable of forming SAM is a fast, simple, low-cost, and effective technique for the development of novel materials especially for the production of electronic devices. The ability to modify its properties by SAM technique has opened up a wide range of applications in electronic and optoelectronic devices.

Keywords: self-assembled monolayer, graphene, surface modification, interface modification, optoelectronic devices

1. Introduction

The development of more efficient, multifunctional, and miniaturized electronic devices requires a continuous research in the area of suitable materials to meet the demand of the modern society. Self-assembled monolayers (SAMs), which are arranged spontaneously on surfaces of the substrate due to the chemical or physical interactions of molecules with a substrate, have gained considerable attention in the past decade. SAMs opened a way toward the miniaturization in microelectronics as well as tailoring the properties of surfaces such as graphene [1–6].

This review focuses on researches committed over the past decade about the modification of graphene via SAM technique. In Section 2, a brief overview of the modification methods of graphene with SAMs was given. In Section 3, the applications of graphene modified by SAMs are summarized and discussed.

2. Modification of graphene with self-assembled monolayers

Graphene is a good candidate material for many applications due to its unique optical, mechanical, and electronic properties [7–10]. Pristine graphene is a zero band gap material, chemically inert, and also insoluble in many organic and inor-ganic solvents [11–13]. Therefore, it cannot be used in many applications. In order to achieve commercial viability and provide to the specific needs of device applica-tions, it is necessary to control the electrical properties of graphene without damag-ing its intrinsic properties. To fully utilize its potential, the surface and electronic structures of graphene have been modified via functionalization (post-synthesis) or doping (during in situ synthesis) methods [14–16]. Substitutional doping of graphene occurs during the synthesis by covalently bonding the chemical groups and thereby transforms carbon hybridization from sp^2 to sp^3 [17–19]. On the other hand, functionalization of graphene is a physical process involving the addition of functional species onto graphene sheets to manipulate van der Waals interactions occurring effectively among the graphene sheets. In the substitution process, elec-trons are exchanged between dopant molecules bonded onto the graphene surface and graphene. Surface charge transfer depends on the relative energy of the high-est occupied molecular orbital (HOMO) or lowest unoccupied molecular orbital (LUMO) of the dopant with respect to the Fermi level of graphene [16, 20–23]. Many different chemical substances have been studied for the functionalization of graphene, and SAM technique is one of them. An ultrathin layer, SAM, can be uniformly formed on the surface by the attachment of the SAM-forming molecule through its functional group as a result of a covalent bond. This technique provides a suitable and flexible way to modify the surface properties of the substrate materi-als to some extent or completely depending on the molecular structure of acting as SAM-forming molecule. Recent reports have also shown that SAMs can significantly change the properties of graphene without sacrificing the intrinsic graphene perfor-mance. In order to modify the electronic properties of graphene by SAM technique, the usage of either electron donating or withdrawing terminal groups has been reported by Yan et al. [24]. As shown in **Figure 1**, the threshold voltage (V_{th}) which corresponds to the neutrality point in graphene field-effect transistor (FET) devices shifted to positive values after the decoration of SiO_2 substrates by using fluori-nated molecules (F-SAMs), indicating p-type doping of the graphene layer. For the graphene FET device made by CH_3-SAMs and H_2N-SAMs, the negative V_{th} shift was observed, indicating p-type FET behaviors. This clearly indicates modification of the electronic properties of graphene which becomes available by suitably adapting the SAM techniques.

SAMs have been widely used to eliminate the Schottky barrier at the materials' interface by modifying the interface between graphene and its support substrate which is creating p-n junctions in graphene [25–27]. Sojoudi et al. used 3-ami-nopropyltriethoxysilane (APTES) and perfluorooctyltriethoxysilane (PFES) for the modification of the interface between transferred CVD-graphene films and its supporting dielectric [28]. Thus, n-type and p-type graphene was created by treating APTES and PFES, respectively. As shown in **Figure 2**, APTES and PFES are patterned on the same dielectric, thus creating a graphene p-n junction. They demonstrated that substrate functionalizing with these SAMs resulted in p-n junc-tions with controlled position and height.

Recent studies have shown that graphene functionalization has been enhanced sig-nificantly by electrochemical control [29–32]. Fermi level of graphene can be shifted and also increased its reactivity toward aggressive chemicals by application of electro-chemical potential. The groups of Treossi and Palermo used electrochemical approach,

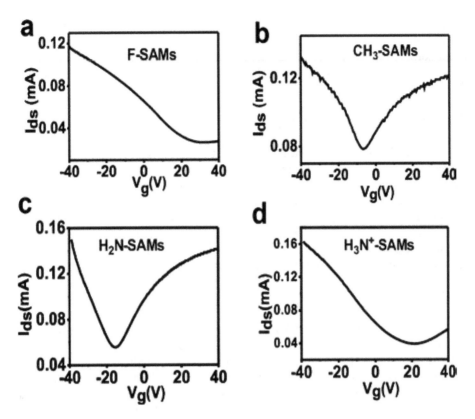

Figure 1.
I_{ds}/V_g characteristics of graphene-based FET devices fabricated on (a) 1H,1H,2H,2H-perfluorooctyltriethoxysilane (F-SAMs), (b) butyltriethoxysilane (CH_3-SAMs), (c) 3-aminopropyltriethoxysilane (H_2N-SAMs), and (d) protonated form produced from the H_2N (H_3N^+-SAMs) (this figure is reprinted from Ref. [24]).

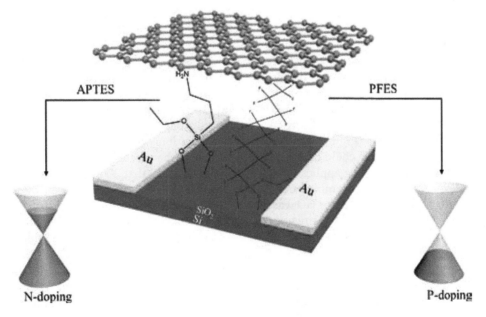

Figure 2.
Schematic representation of graphene p-n junction fabrication.

a fast and simple method, to functionalize graphene [33]. The electrochemical potential allowed good control over the morphology of the 4-docosyloxy-benzenediazonium tetrafluoroborate (DBT) suspended layer, which does not react randomly on the substrate, and also transformed the physisorption into a covalent chemisorption.

3. Applications of graphene/SAM structure

Graphene has well-known electrical properties, which may be utilized in vari-ous applications such as biosensors, environmental sensors, electronic and optical devices, etc. Modification with SAM can result in the change of properties at the interface, such as friction coefficient. The monolayer can also be utilized as a physical spacer, in order to enhance the sensing properties of the layer underneath. Apart from these applications, SAMs can be used in order to synthesize graphene, both by elec-trostatical assembly of graphene flakes onto the surface and by providing the carbon atoms that is required to assemble graphene from their own molecular structure.

3.1 Biosensor

The enhancement effect of SAM is widely used in the field of biosensor and biomedical. Shi et al. formed a SAM from 4-aminothiophenol (4-ATP) solution on gold substrate and placed the prepared substrate in graphene oxide (GO) solution vertically. Coupling reagents ethylene dichloride and N-hydroxysuccinimide (EDC and NHS, respectively) were added to promote the reactions between –NH$_2$ groups of SAM molecules and –COOH groups of GO. After the device fabrication, cyclic voltammetry (CV) and electrochemical impedance spectroscopy (EIS) measurements were carried out, and ultrasensitive electrochemical Hg ion sensing properties were shown [34]. Sun et al. prepared GO and fluorescent SAMs with thrombin aptamer (TBA) fixed onto the outermost layer and used CdTe quantum dots as a fluorescence probe, which allow sensing biomolecules by interfacial energy transfer from CdTe in SAMs to GO. The measured fluorescence spectra showed that the increase of the intensity was related to the concentration of the target molecule (**Figure 3**) and the linear correlation was obtained by using GO which was necessary for defining the device as a sensor [35].

Thakur et al. prepared a graphene-based FET sensor using α-ethyl-tryptamine (AET) to add functionality to the electrodes. This approach enables the electrostatical adsorption of functional groups that are present in GO monolayers, such as carboxylic acid and epoxide, by the amine end of AET. The selectivity and response time showed great results, being able to detect a single *E. coli* cell from a vast amount of sample (**Figure 4**). With the incorporation of suitable probing molecules, this approach could be a useful tool to detect other bacteria as well [36]. For related bacteria sensing applications, Chang et al. reported that 2-aminoethanethiol molecules were self-assembled onto Au electrodes by immersion, and then thermally reduced monolayer of GO was deposited on the electrodes due to electrostatic interactions. After some treatments, the device was incubated in the PBS containing anti-*E. coli* O157 antibodies and introduced as a high-performance FET chemical sensor and biosensor [37].

As a selectivity study, Khan et al. suggested an ITO-based, APTES-modified sensor that contained electrochemically reduced GO. Positively charged amino groups (NH$_3^+$) within APTES have interacted with the negatively charged carboxyl and hydroxyl groups within GO. This electrostatic interaction has allowed deposit of GO onto ITO/APTES electrode. It was observed that the electronic transport is improved with the modification of the ITO surface with SAM, and the addition of

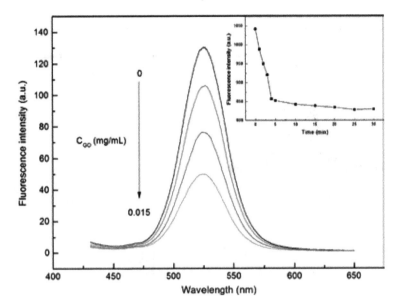

Figure 3.
Fluorescence spectra of sensor at different concentration of tDNA, inset: plot of fluorescence intensity ratio related to concentration with and without GO (figure image reprinted from Ref. [35]).

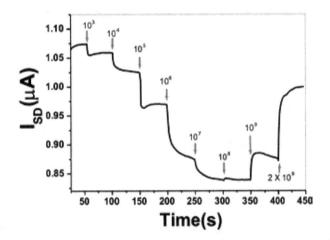

Figure 4.
Real-time detection (V_{ds} = 0.1 V) of E. coli (CFU/mL) in water with the anti-E. coli/AuNPs/Al$_2$O$_3$/rGO biosensor (figure image reprinted from Ref. [36]).

GO onto the ITO/SAM surface increased the conductivity. Therefore, the resulting device had the property of sensing the oxidation peak of tyramine due to selective behavior of electrode toward electrooxidation. The sensor had also stability and selectivity tests, which it stayed stable for 3 weeks (with the loss of <5% activity) and as selective for human urine, commercial milk, and beer sample. Final conclusions ended up with that the device had sensing capabilities for dopamine and ascorbic acid as well, and it is ready for commercial applications [38].

Feng et al. studied a new type of electrochemical aptasensor for highly sensitive detection of adenosine triphosphate (ATP) and adenosine deaminase (ADA) activity. For modification, 6-mercaptohexanol (MCH) was used. Fabricated sensor could sense different molecules with varying concentrations and had amplified electrochemi-cal signals, due to the facilitation of electron transfer of the functionalized graphene (**Figure 5**) [39]. Yeh et al. demonstrated graphene field-effect transistors (G-FETs) for

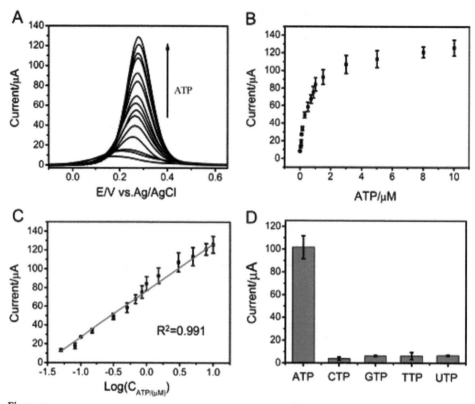

Figure 5.
(A) Differential pulse voltammetry (DPV) responses, (B) current intensities with different ADA active units, (C) a linear relationship between DPV peaks and ADA active units, and (D) specificity of the assay for ADA activity (figure image reprinted from Ref. [39]).

high-mobility applications that were both rigid and flexible. First, graphene was synthesized by CVD method and then transferred onto silicon substrates. The graphene-coated silicone samples were soaked into the solution of hexamethyldisilazane (HMDS) used as SAM-forming molecule for 17 h in a partially covered beaker. The samples were turned into G-FET devices using e-beam lithography and functionalized. As a result, they showed that modification of substrates for graphene electronic devices through patterned SAM arrays of HDMS can enhance the electrical properties of G-FETs by limiting and screening the interactions between graphene and hydrophilic polar groups attached. The HMDS-modified G-FET devices they built could detect cancer biomarker CSPG4 in serum samples at a concentration of as low as 0.01 fM, which is five orders of magnitude more sensitive than a conventional colorimetric assay [40].

Chiu et al. modified a gold electrode using 1-octadecanethiol (ODT) as SAM-forming molecule and then immersed it into GO aqueous solution. After the completion of the coating process, the GO film could be turned into electrochemically reduced GO (ERGO) films. The resulting film showed sensing properties, and the process variables can control the residual oxygen functionalities and conductivity in GO sheets in order to provide tools for sensitive immunoassay detection (**Figure 6**) [41]. Bhardwaj et al. prepared a biological sensor that contained covalently bonded GO array with a layer of SAM formed by 4-aminothiophenol on Au electrodes. After functionalizing the surface with lipase enzyme, the sensor could bind with tributyrin and found to be linearly correlated with concentration between certain limits [42].

In a study of binding mechanism, Wang et al. reported that included CVD-grown graphene, hexagonal boron nitride (hBN), and OTS as SAM-forming molecules. They also used a stamp-like micropatterning process in order to give SAM layer a

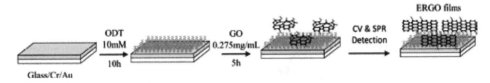

Figure 6.
Fabrication procedure of GO film.

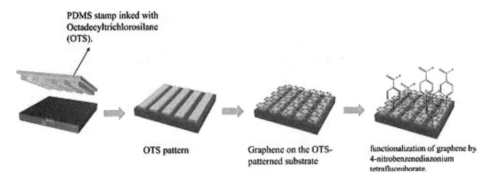

Figure 7.
Illustration of reactivity imprint lithography. PDMS stamp inked with OTS-SAM was used to pattern SiO₂ substrate. Graphene is transferred onto the patterned substrate and reacted with 4-nitrobenzenediazonium tetrafluoroborate.

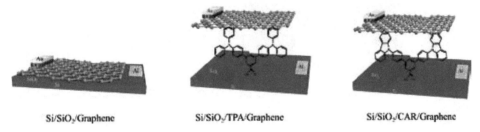

Figure 8.
The structure of devices.

pattern before the coating of the graphene, which affected the binding mechanism of target molecules (**Figure 7**) [43].

3.2 Device

One of the several properties of SAM layers is to alter the working conditions of the bare device. Aydin et al. investigated the properties of a diode that contained graphene, 4″bis(diphenylamino)-1, 1′:3″-terphenyl-5′carboxylic acids (TPA), and 4,4-di-9H-carbazol-9-yl-1,1′:3′1′-terphenyl-5′carboxylic acid (CAR) aromatic SAMs (**Figure 8**). The decrease in the Schottky barrier height and series resistance by forming a suitable interface between Si and graphene is also noted. The graphene forms π-π interaction with aromatic SAMs as well. The theoretical calculations of the study also confirmed that CAR molecules showed better results than TPA, due to the more compatible energy levels [44].

Kim et al. studied the electrical transport characteristics of graphene transistors functionalized with a SAM of cationic molecules. They used 1-aminoethanethiol as SAM molecule onto mechanically exfoliated graphene, and the device is completed after the electrical contact electrodes were fabricated by e-beam lithography (**Figure 9**).

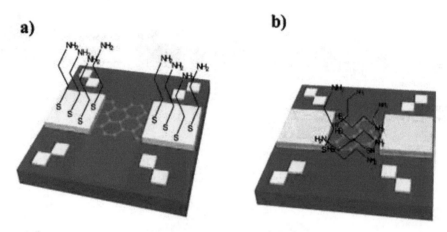

Figure 9.
Schematic illustration of (a) the contact-opened and (b) the channel-opened device.

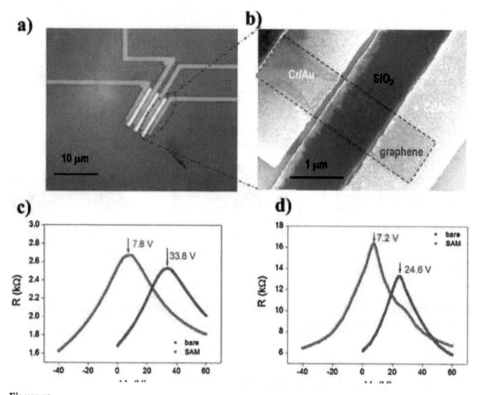

Figure 10.
(a) Optical microscopy (b) SEM images of the contact-opened graphene device. The effects of SAM modification on (c) contact-opened and (d) channel-opened graphene transistor (figure image reprinted from Ref. [45]).

Finally they found that work-function engineering with SAMs results in a larger doping effect than the simple adsorption of amine-containing molecules onto the graphene channels (**Figure 10**) [45]. Jung et al. fabricated graphene hybrid film consisted of graphene/APTES/polyethylene terephthalate (PET) for enhancement of the conductiv-ity. Basically, a CVD-grown graphene film is transferred onto an APTES layer on PET substrate. This process enhanced the conductivity of graphene effectively. As a result of the study, the carrier density of graphene was improved from 9.3×10^{12} to $1.16 \times 10^{13}/cm^2$ with transmittances of 82% and 86%, depending on the process parameters [46].

Gan et al. studied a new method for controlling H- and J-stacking in self-assembly. They used the property of GO and reduced graphene oxide (RGO) in order to control the self-assembly of perylene (Py). This new method suggested new materials for photosensitive applications in optoelectronics and liquid-junction solar cells [47]. Wieghold et al. demonstrated optoelectronic function in almost monolayer molecular architecture. The active layer consisted of a self-assembled terrylene diimide (TDI) derivative dye that formed a bicomponent supramolecular network with melamine. They found photocurrents of 0.5 nA and open-circuit voltages of 270 mV employing 19 mW/cm^2 irradiation intensities at 710 nm. They also estimated an incident photon to current efficiency of 0.6% at 710 nm with a contact area of 9.9×10^2 mm^2, meaning the opening up to the intriguing possibilities in bottom-up optoelectronic device fabrication with molecular resolution [48].

Kang et al. presented a new efficient and stable RGO doping method by employ-ing two different types of alkylsilane compounds to produce p-type or n-type doping of RGO: (tridecafluoro-1,1,2,2,-tetrahydrooctyl)trichlorosilane (FTS) and APTES, respectively (**Figure 11**). The tunable electrical property of the SAM-functionalized RGO was utilized in developing source/drain (S/D) electrodes in bottom contact OFETs. The OFET device performance was improved significantly upon the use of FTS-RGO S/D electrodes [49]. Park et al. devised a method to optimize the performance of OFETs by controlling the work functions of graphene electrodes by functionalizing the surface of SiO$_2$ substrates with N, N′ - ditridecyl - 3,4,9,10-perylenetetracarboxylic

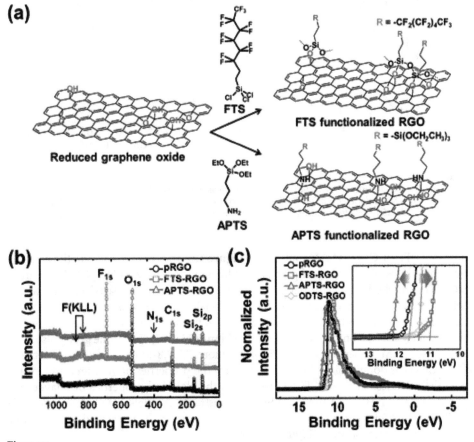

Figure 11.
(a) Representation of SAM functionalization of RGO with FTS and APTES, (b) XPS and (c) UPS spectra of SAM-functionalized RGOs. The inset figure in (c) shows the secondary electron emission region (figure image reprinted from Ref. [49]).

Figure 12.
Schematic illustration of PTCDI-C13 FET device process with patterned graphene onto SAM-modified SiO₂ substrate.

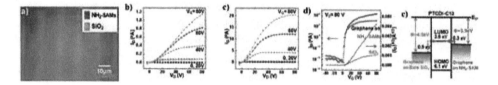

Figure 13.
(a) Optical microscopy images of samples. I-V measurements of (b) graphene S/D electrode containing n-type PTCDI-C13 FETs on pure SiO₂ and (c) on NH₂-SAMs, (d) transfer characteristics of electrodes on different SAM-modified SiO₂. (e) Band diagram structures of samples (figure image reprinted from Ref. [27]).

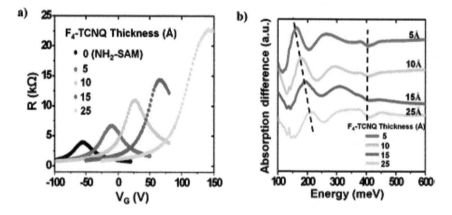

Figure 14.
(a) Current-voltage transfer characteristics of bilayer graphene FETs on a NH₂-SAM-modified SiO₂/Si substrates with various thicknesses of the F4-TCNQ layers, (b) absorption spectra of graphene on NH₂-SAM-modified SiO₂/Si substrates after applying different thickness of F4-TCNQ layers (figure image reprinted from Ref. [50]).

diimide (PTCDI-C13) as SAM molecules (**Figure 12**). In order to facilitate the electron injection, the work function of graphene is lowered to 3.9 eV in NH₂-SAM-modified SiO₂. The method of performance optimization of graphene-based OFETs which utilizes work-function engineering by functionalizing the substrate with SAMs is described and shows ~10 times enhancements in properties regarding the charge carrier mobility and the on-off ratio of OFETs (**Figure 13**) [27]. The same group also used APTES as SAM molecule and 2,3,5,6-tetrafluoro-7,8,8,8-tetracyanoquinodimethane (F4-TCNQ) as doping agent in order to modify the properties of graphene and examined the thickness of the F4-TCNQ related to the electrical properties of bilayer graphene FETs on NH₂-SAM-modified SiO₂/Si substrates. A single-gate bilayer graphene FET with high-current on/off current ratios is successfully fabricated, and tunable bandgap was demonstrated. (**Figure 14**) [50].

Gan et al. discovered a weak organic acid, 3,4,9,10-perylene tetracarboxylic acid (PTCA), and utilized to synthesize a supramolecular nanocomplex containing PTCA-graphene that showed wide voltage window and ultrahigh specific capacitance [51].

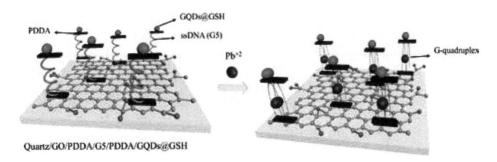

Figure 15.
Schematic diagram of the Pb²⁺ adding into quartz/GO/PDDA/G5/PDDA/GQDs@GSH SAMs.

In terms of environmental sensor, Sun et al. developed a sensor for the detection of Pb^{2+} ions by constructing quartz/GO/PDDA/G5/PDDA/GQDs@GSH system which included GO, glutathione-functionalized graphene quantum dots (GQDs@ GSH), poly(diallydimethylammonium) chloride (PDDA), and G5 (G-rich DNA) (**Figure 15**). The group showed that the detection concentration of the sensor for Pb^{2+} ions is as low as 2.2 nM [52].

3.3 Interface modifier

Pfaffeneder-Kmen et al. used 4-mercaptophenol as SAM molecule in order to pro-duce homogeneous GO coatings. This process allowed the modification of the surface for contact angle to be low enough in order to achieve complete wetting that causes GO suspension to adhere onto the Au surface (**Figure 16a**) [53]. Larisika et al. used APTES as SAM-forming layer on SiO_2 substrates in order to enhance the adsorption of GO before the dip-coating process. The resulting sensor had potential as a high-performance and low-voltage operating graphene device [54]. Kamiya et al. investi-gated protein adsorption to graphene films which modificates the surface of substrates with two different SAMs in aqueous environment, octadecyltrichlorosilane (OTS) and APTES. They employed latex bead projection method for patterning SAMs and observed that the protein adsorption behavior on the graphene flakes is attached to par-tially SAM-modified SiO_2/Si substrates in aqueous environment (**Figure 16b–c**). It was also found that a high-density of agglomerated avidin molecule clusters was formed in the OTS-supported graphene areas, whereas a low-density of large clusters was formed in the SiO_2 and APTES-supported ones. For the high-performance graphene biosensors with small nonspecific adsorption of protein molecules, the suggested technique can control protein adsorption phenomena on graphene surfaces [55].

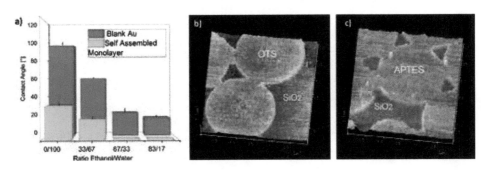

Figure 16.
(a) Contact angle vs. water/ethanol ratio with and without mercaptophenol SAM (figure image reprinted from Ref. [53]). AFM images of (b) an OTS-modified and (c) an APTES-modified SiO_2 substrate (figure image reprinted from Ref. [55]).

Ermakova et al. used the zinc complex of 5,10,15,20-tetra(4-pyridyl)-21H,23H-porphine (ZnTPyP) as an organic promoter for the interfacial adsorption of GO. The ZnTPyP/GO/ZnAc$_2$ bilayer was transferred as whole onto silicon slide by vertical LB method. As conclusion, they informed the stabilizing effect of GO which both anchored the assembled organic layer on the surface. By this result, the final multilayer structures showed large-area uniformity, especially for applications in catalysis, sensors, optoelectronics, etc. [56].

Hui et al. utilized a SAM molecule, 11-mercapto-1-undecanol (11-MU), for the blocking of Au surface, discovering the effect of the electrochemical kinetics of external-sphere redox mediators. For this purpose, metal electrodes were buried in the subsurface of continuous double-layer graphene electrodes. Modified Au tip exhibited a substantial decrease in its electron transfer kinetics [57].

In terms of mechanical modification, Bai et al. used APTES-SAM molecules between Si substrate and ceria/GO composite films in order to alter the mechanical properties of the final film. The friction coefficient was reduced drastically, and the anti-wear lifetime was longer than the GO films (**Figure 17**) [58]. Li et al. investigated and compared three different structures: titanium substrate, APTES-SAM, and GO-APTES nanolayer. The modifications changed the wettability, adhesion, and friction forces of the final structures (**Figure 18**) [59].

Li et al. used APTES as an intermediate coupling agent for chemisorption of RGO sheets on titanium alloy substrate. The results showed that the prepared films contain chemisorptive bonds on the substrate. Tribological results indicated that the prepared APTES-RGO film decreased the friction coefficient while improving antifriction properties of titanium alloy under dry friction (**Figure 19a**) [60]. Ou et al. utilized two different SAM layers including APTES and OTS in order to study lubrication coating properties. First, an APTES-SAM with amine (—NH$_2$) outer

Figure 17.
SEM images of the wear scars on the steel balls sliding against (a) APTES-GO and (b) CeO$_2$/GO composite film. (c) Variation of kinetic friction coefficient with time for different samples under an applied load of 2 N and a constant speed of 10 mm/s (1 Hz) (figure image reprinted from Ref. [58]).

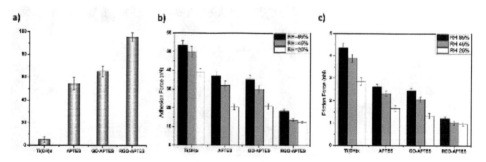

Figure 18.
(a) Water contact angle results of Ti substrate and prepared films; (b) adhesion force and (c) friction force with a normal load of 70 nN at ambient conditions at different relative humidity (figure image reprinted from Ref. [59]).

groups was covalently anchored onto the Si wafer. GO sheets were chemo-grafted onto the APTES-SAM surface. Finally, an OTS outer layer was assembled onto the GO surface (**Figure 19b**). In order to increase the hydrophobicity of the film, an outer layer of OTS-SAM is introduced, which reduced the friction and boosted the anti-wear life [61].

Lee et al. used four different SAMs, APTES, (3-aminopropyl)trime-thoxysilane (APTMS), (3-glycidyloxypropyl)trimethoxysilane (GPTMS), and triethoxym-ethylsilane (MTES), in order to modify GO and silane-functionalized GO (sGO). The samples were prepared using epoxy and tested mechanically. The results showed that sGO, especially those containing amine functional groups, can strengthen the interfacial bonding between the carbon fibers and epoxy adhesive (**Figure 20**) [62]. Liu et al. used negatively charged poly(sodium 4-styrenesulfonate) (PSS)-mediated graphene sheets (PSS-GS) and the posi-tively charged polyethyleneimine (PEI) repeatedly layer by layer, in order to obtain a multilayered film. Both the layer thickness and layer numbers were

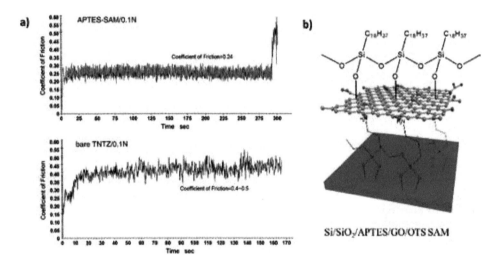

Figure 19.
(a) Tribology results of varied materials on TNTZ surface in contact with Si_3N_4 balls under dry friction, (top) APTES-SAMs, 0.1 N applied load; (bottom) bare TNTZ, 0.1 N (figure image reprinted from Ref. [60]). (b) Schematic illustration of APTES-GO-OTS film structure on a Si wafer.

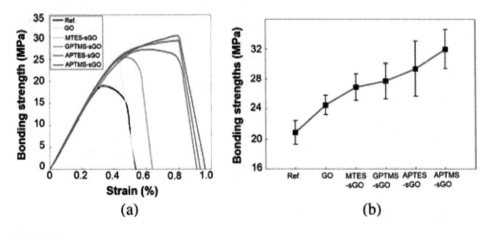

Figure 20.
(a) Bonding strength-strain relationships with the addition of GO or the sGOs in the carbon fiber/epoxy composite, (b) average bonding strength of the sample (figure image reprinted from Ref. [62]).

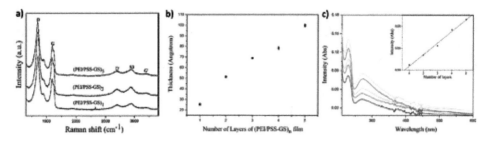

Figure 21.
(a) Raman spectra of sample films; (b) graph showing the correlation between the layer numbers of sample and ellipsometric thicknesses; (c) UV-vis spectroscopy results of samples that are assembled on a quartz slide via LBL method. Inset shows the correlation between the absorbance value at 270 nm and the layer number (n) (figure image reprinted from Ref. [63]).

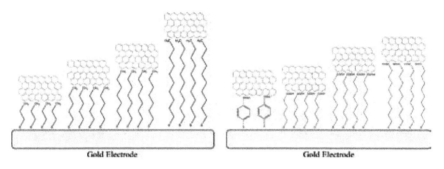

Figure 22.
Schematic representation of CRGOs' self-assembly process on to the gold electrode.

changed in order to determine the relation. Their good tribological properties under the given testing conditions suggested that the use as a low-friction and anti-wear coating layer is possible (**Figure 21**) [63].

3.4 Spacer

Kong et al. reported the influence of chemically reduced graphene oxide sheets (CRGOs) on the electrochemical performance through methyl or carboxylic acid-terminated SAMs (**Figure 22**). Modified gold electrode and immobilization of the CRGOs on a SAM-treated surface showed effective enhancement of the hetero-geneous electron transfer (ET) of the SAM due to the tunneling effect. They also reported that the kinetics of electron transfer activity between the CRGOs/SAM/Au electrode and redox species in the solution was attributed to charge transfer being confined to CRGOs with different interactions with —CH_3 and —COOH terminated thiols, which can influence the electron transfer efficiency and the rate of charge transfer due to different electron transfer pathways [64].

Margapoti et al. reported the experimental studies of resonance oscillation in the current density of a SAM-graphene hybrid system. They used Au surface function-alized with a mixture of 4-(1-mercapto-6-hexyloxy) azobenzene as SAM treatment and 6-mercapto-1-hexanol as spacer molecules. A reversible change in conductiv-ity was observed after transforming the molecular configuration from *trans* to *cis* (**Figure 23**) [65].

3.5 Graphene synthesis by SAM

It is also quite possible to synthesize high-quality graphene by using SAM technique as assembling surfaces. Xie et al. reported a simple and environmentally

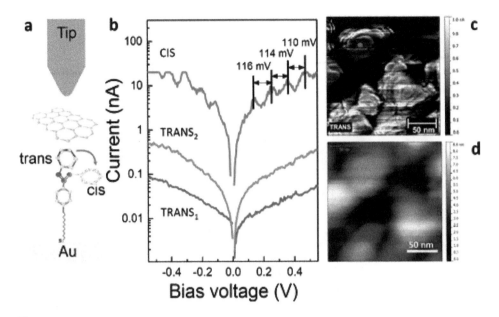

Figure 23.
(a) Measurement configurations and schematic representation of graphene/mSAM/Au hybrid with the azobenzene molecule in trans- and cis-configurations. (b) Typical I-V characteristics in trans-configuration before illumination (blue line) and in cis-configurations, which is controlled via UV-light (red line) and trans-configuration (green line) following white light exposure. (c) Topographical image and (d) current topography with the mSAM in the trans-configuration (figure image reprinted from Ref. [65]).

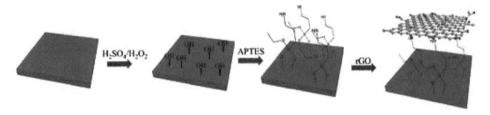

Figure 24.
Schematic representation for self-assembly of CRGO monolayer on APTES/Si/SiO₂ surface.

friendly approach for synthesizing a composite of graphene/gold nanoparticles (3DG/Au NPs) in one step. The selected molecule for SAM treatment is 4-amino-thiophenol. They immersed the SAM-coated glass substrate into the GO-HAuCL₄ solution, and the resulting sample showed catalytic reduction properties [66].

Yin et al. successfully bonded monolayer of CRGO nanosheets on Si sub-strates chemically and investigated their possible applications in Raman scatter-ing (**Figure 24**). The assembly of large-scale and uniform graphene sheet was performed by the electrostatic absorption between the —NH₂ groups of APTES that was coated onto the Si surface and residual groups on RGO. They also concluded that the mildly RGO substrate has the optimum graphene-enhanced Raman scattering (GERS) performance among all the CRGO substrates. The π-π stacking and the residual polarized oxygen groups on CRGO surface were mainly responsible for the excellent GERS effect of mildly RGO substrate [67].

Jing et al. reported a facile approach to the direct synthesis of graphene sheets based on the SAM technique. They used a thiophene-terminated silane molecule, triethoxy-(6-(thiophen-3-yloxy)-hexyl)silane (TEHS), as carbon source, and by heating the cross-linked polythiophene SAM up to 1000°C under high vacuum, single-layered or few-layered graphene sheets were successfully prepared on the

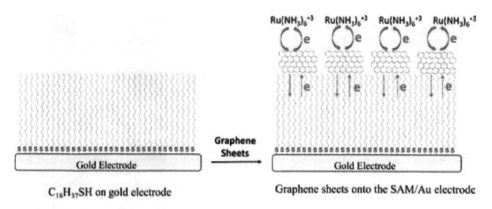

Figure 25.
Illustration of the fabrication of a graphene-/SAM-modified gold electrode and electron transfer of the mechanism on the graphene-modified electrode.

dielectric silicon oxide substrate [68]. Xie et al. investigated the electrochemical behaviors of graphene sheets attached to a SAM (n-octadecyl mercaptan) on a gold electrode (**Figure 25**). Their studies revealed that the heterogeneous electron transfer blocked by the SAM layer can be restored by graphene sheets, and the graphene-/SAM-modified Au electrode had a smaller interfacial capacitance, as compared with that of a bare one [69].

Three innovative methods are introduced by Nayak et al. for functionalizing bioinert ceramic. These methods include immobilizing rGO onto the surface using appropriate SAMs. Among these methods, activated —COOH groups of rGO and —NH_2 functionalized zirconia shows the most effective immobilization. They also concluded that further immobilization makes it possible to customize the surface properties for the desired application and could be easily performed on graphene-modified ceramics [70].

4. Conclusions

Following Geim and Novoselov study about graphene, it has been the most exciting and promising material discovered in the last few years. Due to its outstanding properties, it has enormous potential to develop many applications ranging from nanoelectronics to nanobiotechnology [71, 72]. However, tuning the physical and chemical properties of graphene, which possess high sheet resistance and zero bandgap, is crucial to the realization of graphene-based technologies. Furthermore, adsorption of graphene onto a solid surface is important for the device applications since it provides good charge transfer. Functionalization of the surface with SAMs is useful for controlling the electronic properties of the graphene layer and improving adsorption, paving the way to develop commercial applications.

In recent years, the interaction between SAMs with different chemical properties and graphene has been investigated by many research groups. For instance, SAMs with thiol group is most widely studied for electronic applications due to having delocalized π-electrons. These SAMs can also modify electrodes for electrochemical sensor applications. For example, SAM-decorated Au probes can electrostatically adsorb graphene derivatives, both functionalizing the sensor and improving the sensing ability. These researches demonstrate that SAMs are a utility for improving the performance of experimental devices. It is considered widely that overcoming the difficulties in the production of large-area graphene sheets may allow develop-ing commercial applications.

In this chapter, we have demonstrated the functionalization, modification, and synthesis of graphene by SAMs and reviewed different applications. This chapter also demonstrates how to utilize the SAM-modified graphene and that SAM modifi-cation can improve the possibility of its usage in many applications.

Author details

Gulsum Ersu[1], Yenal Gokpek[1], Mustafa Can[2], Ceylan Zafer[3] and Serafettin Demic[1]*

1 Department of Material Science and Engineering, Faculty of Engineering and Architecture, Izmir Katip Celebi University, Izmir, Turkey

2 Department of Engineering Sciences, Faculty of Engineering and Architecture, Izmir Katip Celebi University, Izmir, Turkey

3 Ege University, Institute of Solar Energy, İzmir, Turkey

*Address all correspondence to: serafettin.demic@ikc.edu.tr

References

[1] Heath JR. Molecular electronics. Annual Review of Materials Research. 2009;**39**(1):1-23. DOI: 10.1146/annurev-matsci-082908-145401

[2] Newton L, Slater T, Clark N, Vijayaraghavan A. Self assembled monolayers (SAMs) on metallic surfaces (gold and graphene) for electronic applications. Journal of Materials Chemistry C. 2013;**1**(3):376-393. DOI: 10.1039/C2TC00146B

[3] Devanarayanan VP, Manjuladevi V, Gupta RK. Interaction of graphene with self assembled monolayers. Macromolecular Symposia.2015; **357**(1):23-29. DOI: 10.1002/masy.201400180

[4] Tu Q, Kim HS, Oweida TJ, Parlak Z, Yingling YG, Zauscher S. Interfacial mechanical properties of graphene on self-assembled monolayers: Experiments and simulations. ACS Applied Materials & Interfaces. 2017;**9**(11): 10203-10213. DOI: 10.1021/acsami.6b16593

[5] Schwartz DK. Mechanisms and kinetics of self-assembled monolayer formation. Annual Review of Physical Chemistry. 2001;**52**(1):107-137. DOI: 10.1146/annurev.physchem.52.1.107

[6] Lee WH, Park J, Kim Y, Kim KS, Hong BH, Cho K. Control of graphene field-effect transistors by interfacial hydrophobic self-assembled monolayers. Advanced Materials. 2011;**23**(30):3460-3464. DOI: 10.1002/adma.201101340

[7] Geim AK, Novoselov KS. The rise of graphene. Nature Materials. 2007;**6**:183. DOI: 10.1038/nmat1849

[8] Bolotin KI, Sikes KJ, Jiang Z, Klima M, Fudenberg G, Hone J, et al. Ultrahigh electron mobility in suspended graphene. Solid State Communications. 2008;**146**(9):351-355. DOI: 10.1016/j.ssc.2008.02.024

[9] Choi W, Lahiri I, Seelaboyina R, Kang YS. Synthesis of graphene and its applications: A review. Critical Reviews in Solid State and Materials Sciences. 2010;**35**(1):52-71. DOI: 10.1080/10408430903505036

[10] Kim K, Choi J-Y, Kim T, Cho S-H, Chung H-J. A role for graphene in silicon-based semiconductor devices. Nature. 2011;**479**:338. DOI: 10.1038/nature10680

[11] Novoselov KS, Fal'ko VI, Colombo L, Gellert PR, Schwab MG, Kim K. A roadmap for graphene. Nature. 2012;**490**:192. DOI: 10.1038/nature11458

[12] Smerieri M, Celasco E, Carraro G, Lusuan A, Pal J, Bracco G, et al. Enhanced chemical reactivity of pristine graphene interacting strongly with a substrate: Chemisorbed carbon monoxide on graphene/nickel(1 1 1). ChemCatChem. 2015;**7**(15):2328-2331. DOI: 10.1002/cctc.201500279

[13] Wang X, Shi G. An introduction to the chemistry of graphene. Physical Chemistry Chemical Physics. 2015;**17**(43):28484-28504. DOI: 10.1039/C5CP05212B

[14] Liu H, Liu Y, Zhu D. Chemical doping of graphene. Journal of Materials Chemistry. 2011;**21**(10):3335-3345. DOI: 10.1039/C0JM02922J

[15] Wang X, Sun G, Routh P, Kim D-H, Huang W, Chen P. Heteroatom-doped graphene materials: Syntheses, properties and applications. Chemical Society Reviews. 2014;**43**(20):7067-7098. DOI: 10.1039/C4CS00141A

[16] Bong JH, Sul O, Yoon A, Choi S-Y, Cho BJ. Facile graphene n-doping

by wet chemical treatment for electronic applications. Nanoscale. 2014;**6**(15):8503-8508. DOI: 10.1039/C4NR01160K

[17] Georgakilas V, Otyepka M, Bourlinos AB, Chandra V, Kim N, Kemp KC, et al. Functionalization of graphene: Covalent and non-covalent approaches, derivatives and applications. Chemical Reviews. 2012;**112**(11): 6156-6214. DOI: 10.1021/cr3000412

[18] Sarkar S, Bekyarova E, Haddon RC. Covalent chemistry in graphene electronics. Materials Today. 2012;**15**(6):276-285. DOI: 10.1016/S1369-7021(12)70118-9

[19] Johns JE, Hersam MC. Atomic covalent functionalization of graphene. Accounts of Chemical Research. 2013;**46**(1):77-86. DOI: 10.1021/ar300143e

[20] Kaur P, Shin M-S, Sharma N, Kaur N, Joshi A, Chae S-R, et al. Noncovalent functionalization of graphene with poly(diallyldimethylammonium) chloride: Effect of a nonionic surfactant. International Journal of Hydrogen Energy. 2015;**40**(3): 1541-1547. DOI: 10.1016/j.ijhydene.2014.11.068

[21] Wei P, Liu N, Lee HR, Adijanto E, Ci L, Naab BD, et al. Tuning the dirac point in CVD-grown graphene through solution processed n-type doping with 2 - (2 - methoxyphenyl)- 1,3-dimethyl-2,3-dihydro-1H-benzoimidazole. Nano Letters. 2013;**13**(5):1890-1897. DOI: 10.1021/nl303410g

[22] Cai B, Zhang S, Yan Z, Zeng H. Noncovalent molecular doping of two-dimensional materials. ChemNanoMat. 2015;**1**(8):542-557. DOI: 10.1002/cnma.201500102

[23] Mao HY, Lu YH, Lin JD, Zhong S, Wee ATS, Chen W. Manipulating the electronic and chemical properties of graphene via molecular functionalization. Progress in Surface Science. 2013;**88**(2):132-159. DOI: 10.1016/j.progsurf.2013.02.001

[24] Yan Z, Sun Z, Lu W, Yao J, Zhu Y, Tour JM. Controlled modulation of electronic properties of graphene by self-assembled monolayers on SiO_2 substrates. ACS Nano. 2011;**5**(2):1535-1540. DOI: 10.1021/nn1034845

[25] Ito Y, Virkar AA, Mannsfeld S, Oh JH, Toney M, Locklin J, et al. Crystalline ultrasmooth self-assembled monolayers of alkylsilanes for organic field-effect transistors. Journal of the American Chemical Society. 2009;**131**(26): 9396-9404. DOI: 10.1021/ja9029957

[26] Liu Z, Bol AA, Haensch W. Large-scale graphene transistors with enhanced performance and reliability based on interface engineering by phenylsilane self-assembled monolayers. Nano Letters. 2011;**11**(2):523-528. DOI: 10.1021/nl1033842

[27] Park J, Lee WH, Huh S, Sim SH, Kim SB, Cho K, et al. Work-function engineering of graphene electrodes by self-assembled monolayers for high-performance organic field-effect transistors. The Journal of Physical Chemistry Letters. 2011;**2**(8):841-845. DOI: 10.1021/jz200265w

[28] Sojoudi H, Baltazar J, Tolbert LM, Henderson CL, Graham S. Creating graphene p-n junctions using self-assembled monolayers. ACS Applied Materials & Interfaces. 2012;**4**(9):4781-4786. DOI: 10.1021/am301138v

[29] Xia ZY, Giambastiani G, Christodoulou C, Nardi MV, Koch N, Treossi E, et al. Synergic exfoliation of graphene with organic molecules and inorganic ions for the electrochemical production of flexible electrodes. ChemPlusChem. 2014;**79**(3):439-446. DOI: 10.1002/cplu.201300375

[30] Cooper AJ, Wilson NR, Kinloch IA, Dryfe RAW. Single stage electrochemical exfoliation method for the production of few-layer graphene via intercalation of tetraalkylammonium cations. Carbon. 2014;**66**:340-350. DOI: 10.1016/j.carbon.2013.09.009

[31] Chakrabarti MH, Low CTJ, Brandon NP, Yufit V, Hashim MA, Irfan MF, et al. Progress in the electrochemical modification of graphene-based materials and their applications. Electrochimica Acta. 2013;**107**:425-440. DOI: 10.1016/j. electacta.2013.06.030

[32] Sundaram RS, Gómez-Navarro C, Balasubramanian K, Burghard M, Kern K. Electrochemical modification of graphene. Advanced Materials. 2008;**20**(16):3050-3053. DOI: 10.1002/adma.200800198

[33] Xia Z, Leonardi F, Gobbi M, Liu Y, Bellani V, Liscio A, et al. Electrochemical functionalization of graphene at the nanoscale with self-assembling diazonium salts. ACS Nano. 2016;**10**(7):7125-7134. DOI: 10.1021/acsnano.6b03278

[34] Shi L, Wang Y, Ding S, Chu Z, Yin Y, Jiang D, et al. A facile and green strategy for preparing newly-designed 3D graphene/gold film and its application in highly efficient electrochemical mercury assay. Biosensors and Bioelectronics. 2017;**89**: 871-879. DOI: 10.1016/j.bios.2016.09.104

[35] Sun X, Liu B, Yang C, Li C. An extremely sensitive aptasensor based on interfacial energy transfer between QDS SAMs and GO. Spectrochimica Acta Part A: Molecular and Biomolecular Spectroscopy. 2014;**131**:288-293. DOI: 10.1016/j.saa.2014.04.093

[36] Thakur B, Zhou G, Chang J, Pu H, Jin B, Sui X, et al. Rapid detection of single *E. coli* bacteria using a graphene-based field-effect transistor device. Biosensors and Bioelectronics. 2018;**110**:16-22. DOI: 10.1016/j. bios.2018.03.014

[37] Chang J, Mao S, Zhang Y, Cui S, Zhou G, Wu X, et al. Ultrasonic-assisted self-assembly of monolayer graphene oxide for rapid detection of *Escherichia coli* bacteria. Nanoscale. 2013;**5**(9): 3620-3626. DOI: 10.1039/C3NR00141E

[38] Khan M, Liu X, Zhu J, Ma F, Hu W, Liu X. Electrochemical detection of tyramine with ITO/APTES/ErGO electrode and its application in real sample analysis. Biosensors and Bioelectronics. 2018;**108**:76-81. DOI: 10.1016/j.bios.2018.02.042

[39] Feng L, Zhang Z, Ren J, Qu X. Functionalized graphene as sensitive electrochemical label in target-dependent linkage of split aptasensor for dual detection. Biosensors and Bioelectronics. 2014;**62**:52-58. DOI: 10.1016/j.bios.2014.06.008

[40] Yeh C-H, Kumar V, Moyano DR, Wen S-H, Parashar V, Hsiao S-H, et al. High-performance and high-sensitivity applications of graphene transistors with self-assembled monolayers. Biosensors and Bioelectronics. 2016; **77**: 1008-1015. DOI: 10.1016/j.bios.2015.10.078

[41] Chiu N-F, Yang C-D, Chen C-C, Kuo C-T. Stepwise control of reduction of graphene oxide and quantitative real-time evaluation of residual oxygen content using EC-SPR for a label-free electrochemical immunosensor. Sensors and Actuators B: Chemical. 2018;**258**:981-990. DOI: 10.1016/j. snb.2017.11.187

[42] Bhardwaj SK, Basu T. Study on binding phenomenon of lipase enzyme with tributyrin on the surface of graphene oxide array using surface plasmon resonance. Thin Solid Films. 2018;**645**:10-18. DOI: 10.1016/j. tsf.2017.10.021

[43] Wang QH, Jin Z, Kim KK, Hilmer AJ, Paulus GL, Shih C-J, et al. Understanding and controlling the substrate effect on graphene electron-transfer chemistry via reactivity imprint lithography. Nature Chemistry. 2012;**4**(9):724. DOI: 10.1038/nchem.1421

[44] Aydin H, Bacaksiz C, Yagmurcukardes N, Karakaya C, Mermer O, Can M, et al. Experimental and computational investigation of graphene/SAMs/n-Si Schottky diodes. Applied Surface Science. 2018;**428**:1010- 1017. DOI: 10.1016/j.apsusc.2017.09.204

[45] Kim B-K, Jeon E-K, Lee J-O, Kim J-J. The effects of cationic molecules on graphene transistors. Synthetic Metals. 2013;**181**:52-55. DOI: 10.1016/j. synthmet.2013.08.013

[46] Jung D, Ko Y-H, Cho J, Adhikari PD, Lee SI, Kim Y, et al. Transparent and flexible conducting hybrid film combined with 3-aminopropyltriethoxysilane-coated polymer and graphene. Applied Surface Science. 2015;**357**:287-292. DOI: 10.1016/j.apsusc. 2015.08.139

[47] Gan S, Zhong L, Engelbrekt C, Zhang J, Han D, Ulstrup J, et al. Graphene controlled H-and J-stacking of perylene dyes into highly stable supramolecular nanostructures for enhanced photocurrent generation. Nanoscale. 2014;**6**(18):10516-10523. DOI: 10.1039/C4NR02308K

[48] Wieghold S, Li J, Simon P, Krause M, Avlasevich Y, Li C, et al. Photoresponse of supramolecular self-assembled networks on graphene-diamond interfaces. Nature Communications. 2016;**7**:10700. DOI: 10.1038/ncomms10700

[49] Kang B, Lim S, Lee WH, Jo SB, Cho K. Work-function-tuned reduced graphene oxide via direct surface functionalization as source/drain electrodes in bottom-contact organic transistors. Advanced Materials. 2013;**25**(41):5856-5862. DOI: 10.1002/adma.201302358

[50] Park J, Jo SB, Yu YJ, Kim Y, Yang JW, Lee WH, et al. Single-gate bandgap opening of bilayer graphene by dual molecular doping. Advanced Materials. 2012;**24**(3):407-411. DOI: 10.1002/adma.201103411

[51] Gan S, Zhong L, Gao L, Han D, Niu L. Electrochemically driven surface-confined acid/base reaction for an ultrafast H^+ supercapacitor. Journal of the American Chemical Society. 2016;**138**(5):1490-1493. DOI: 10.1021/jacs.5b12272

[52] Sun X, Peng Y, Lin Y, Cai L, Li F, Li B. G-quadruplex formation enhancing energy transfer in self-assembled multilayers and fluorescence recognize for Pb^{2+} ions. Sensors and Actuators B: Chemical. 2018;**255**: 2121-2125. DOI: 10.1016/j.snb. 2017.09. 004

[53] Pfaffeneder-Kmen M, Casas IF, Naghilou A, Trettenhahn G, Kautek W. A multivariate curve resolution evaluation of an in-situ ATR-FTIR spectroscopy investigation of the electrochemical reduction of graphene oxide. Electrochimica Acta. 2017;**255**:160-167. DOI: 10.1016/j. electacta.2017.09.124

[54] Larisika M, Huang J, Tok A, Knoll W, Nowak C. An improved synthesis route to graphene for molecular sensor applications. Materials Chemistry and Physics. 2012;**136**(2-3):304-308. DOI: 10.1016/j. matchemphys.2012.08.003

[55] Kamiya Y, Yamazaki K, Ogino T. Protein adsorption to graphene surfaces controlled by chemical modification of the substrate surfaces. Journal of Colloid and Interface Science. 2014;**431**:77-81. DOI: 10.1016/j. jcis.2014.06.023

[56] Ermakova EV, Ezhov AA, Baranchikov AE, Gorbunova YG, Kalinina MA, Arslanov VV. Interfacial self-assembly of functional bilayer templates comprising porphyrin arrays and graphene oxide. Journal of Colloid and Interface Science. 2018;**530**:521-531. DOI: 10.1016/j.jcis.2018.06.086

[57] Hui J, Zhou X, Bhargava R, Chinderle A, Zhang J, Rodríguez-López J. Kinetic modulation of outer-sphere electron transfer reactions on graphene electrode with a sub-surface metal substrate. Electrochimica Acta. 2016;**211**:1016-1023. DOI: 10.1016/j.electacta.2016.06.134

[58] Bai G, Wang J, Yang Z, Wang H, Wang Z, Yang S. Self-assembly of ceria/graphene oxide composite films with ultra-long antiwear lifetime under a high applied load. Carbon. 2015;**84**:197- 206. DOI: 10.1016/j.carbon.2014.11.063

[59] Li PF, Zhou H, Cheng X-H. Nano/micro tribological behaviors of a self-assembled graphene oxide nanolayer on Ti/titanium alloy substrates. Applied Surface Science. 2013;**285**: 937-944. DOI: 10.1016/j.apsusc.2013.09.019

[60] Li PF, Xu Y, Cheng X-H. Chemisorption of thermal reduced graphene oxide nano-layer film on TNTZ surface and its tribological behavior. Surface and Coating Technology. 2013;**232**:331-339. DOI: 10.1016/j.surfcoat.2013.05.030

[61] Ou J, Wang Y, Wang J, Liu S, Li Z, Yang S. Self-assembly of octadecyltrichlorosilane on graphene oxide and the tribological performances of the resultant film. The Journal of Physical Chemistry C. 2011;**115** (20):10080-10086. DOI: 10.1021/jp200 597k

[62] Lee CY, Bae J-H, Kim T-Y, Chang S-H, Kim SY. Using silane-functionalized graphene oxides for enhancing the interfacial bonding strength of carbon/epoxy composites. Composites Part A: Applied Science and Manufacturing. 2015;**75**:11-17. DOI:10.1016/j.compositesa.2015.04.013

[63] Liu S, Ou J, Li Z, Yang S, Wang J. Layer-by-layer assembly and tribological property of multilayer ultrathin films constructed by modified graphene sheets and polyethyleneimine. Applied Surface Science. 2012;**258**(7):2231-2236. DOI: 10.1016/j.apsusc.2011.09.011

[64] Kong N, Vaka M, Nam ND, Barrow CJ, Liu J, Conlan XA, et al. Controllable graphene oxide mediated efficient electron transfer pathways across self-assembly monolayers: A new class of graphene based electrodes. Electrochimica Acta. 2016;**210**:539-547. DOI: 10.1016/j.electacta.2016.05.143

[65] Margapoti E, Strobel P, Asmar MM, Seifert M, Li J, Sachsenhauser M, et al. Emergence of photoswitchable states in a graphene-azobenzene-Au platform. Nano Letters. 2014;**14**(12): 6823-6827. DOI: 10.1021/nl503681z

[66] Xie J, Yang X, Xu X. Wet chemical method for synthesizing 3D graphene/gold nanocomposite: Catalytic reduction of methylene blue. Physica E: Low-dimensional Systems and Nanostructures. 2017;**88**:201-205. DOI: 10.10 16/j.physe.2016.11.016

[67] Yin F, Wu S, Wang Y, Wu L, Yuan P, Wang X. Self-assembly of mildly reduced graphene oxide monolayer for enhanced Raman scattering. Journal of Solid State Chemistry. 2016;**237**:57-63. DOI: 10.1016/j.jssc.2016.01.015

[68] Jing H, Min M, Seo S, Lu B, Yoon Y, Lee SM, et al. Non-metal catalytic synthesis of graphene from a polythiophene monolayer on silicon dioxide. Carbon. 2015;**86**:272-278. DOI: 10.1016/j.carbon.2015.01.044

[69] Xie X, Zhao K, Xu X, Zhao W, Liu S, Zhu Z, et al. Study of heterogeneous

electron transfer on the graphene/self-assembled monolayer modified gold electrode by electrochemical approaches. The Journal of Physical Chemistry C. 2010;**114**(33):14243-14250. DOI: 10.1021/jp102446w

[70] Nayak GS, Zybala R, Kozinski R, Woluntarski M, Telle R, Schickle K. Immobilization of reduced graphene oxide nano-flakes on inert ceramic surfaces using self-assembled monolayer technique. Materials Letters. 2018;**225**:109-112. DOI: 10.1016/j.matlet.2018.05.004

[71] Liu J, Tang J, Gooding JJ. Strategies for chemical modification of graphene and applications of chemically modified graphene. Journal of Materials Chemistry. 2012;**22**(25):12435-12452. DOI: 10.1039/C2JM31218B

[72] Lonkar SP, Deshmukh YS, Abdala AA. Recent advances in chemical modifications of graphene. Nano Research. 2015;**8**(4):1039-1074. DOI: 10.1007/s12274-014-0622-9

7

Excitons in Two-Dimensional Materials

Xiaoyang Zheng and Xian Zhang

Abstract

Because of the reduced dielectric screening and enhanced Coulomb interactions, two-dimensional (2D) materials like phosphorene and transition metal dichalcogenides (TMDs) exhibit strong excitonic effects, resulting in fascinating many-particle phenomena covering both intralayer and interlayer excitons. Their intrinsic band gaps and strong excitonic emissions allow the possibility to tune the inherent optical, electrical, and optoelectronic properties of 2D materials via a variety of external stimuli, making them potential candidates for novel optoelectronic applications. In this review, we summarize exciton physics and devices in 2D semiconductors and insulators, especially in phosphorene, TMDs, and their van der Waals heterostructures (vdWHs). In the first part, we discuss the remarkably versatile excitonic landscape, including bright and dark excitons, trions, biexcitons, and interlayer excitons. In the second part, we examine common control methods to tune excitonic effects via electrical, magnetic, optical, and mechanical means. In the next stage, we provide recent advances on the optoelectronic device applica-tions, such as electroluminescent devices, photovoltaic solar cells, and photode-tectors. We conclude with a brief discussion on their potential to exploit vdWHs toward unique exciton physics and devices.

Keywords: excitons, two-dimensional materials, semiconductors, heterostructures, optoelectronics

1. Introduction

Since the first 'modern' 2D material, monolayer graphene, was mechanically exfoliated in 2004 [1], the family of 2D materials has been extensively flourishing, covering insulators, semiconductors, semimetals, metals, and superconductors (**Figure 1**). In addition to semimetal graphene, other actively researched 2D materials include wide-bandgap insulator hexagonal boron nitride (hBN) [2], direct bandgap semiconductor phosphorene [3], Xenes (e.g., Monolayers of silicon (silicene), germanium (germanene) and tin (stanene)) [4], and transition metal dichalcogenides (TMDs) with the chemical formula MX_2 (M: transition metal; X: chalcogen) [5]. Compared with bulk materials, 2D materials exhibit some unparallel characteristics: removal of van der Waals interactions, an increase in the ratio of surface area-to-volume, and confinement of electrons in a plane. The change in properties, caused by a reduction in the dimensionality of 2D materials, makes them becoming the promising candidates for next-generation electronics and optoelectronics [6–8].

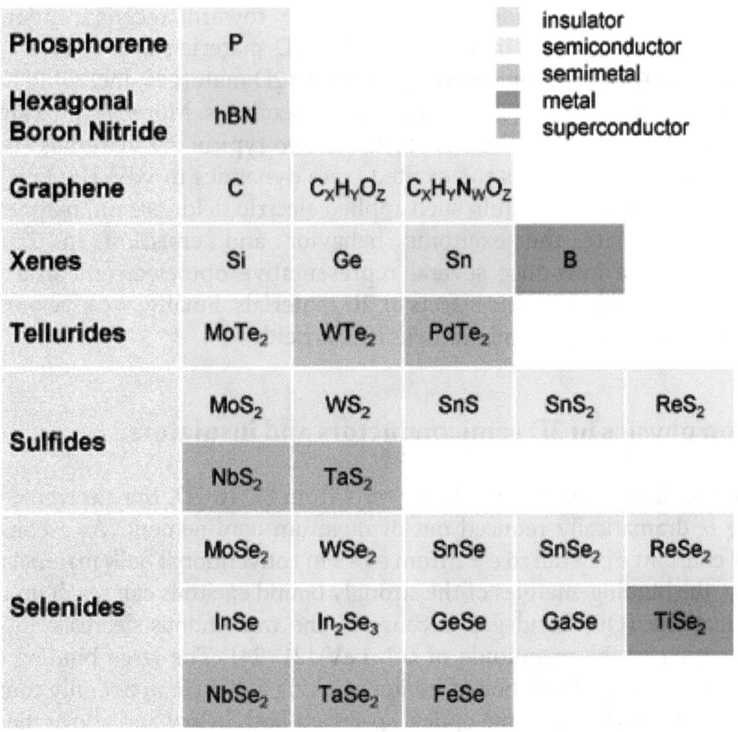

Phosphorene	P			
Hexagonal Boron Nitride	hBN			
Graphene	C	$C_XH_YO_Z$	$C_XH_YN_WO_Z$	
Xenes	Si	Ge	Sn	B

Tellurides	$MoTe_2$	WTe_2	$PdTe_2$		
Sulfides	MoS_2	WS_2	SnS	SnS_2	ReS_2
	NbS_2	TaS_2			
Selenides	$MoSe_2$	WSe_2	SnSe	$SnSe_2$	$ReSe_2$
	InSe	In_2Se_3	GeSe	GaSe	$TiSe_2$
	$NbSe_2$	$TaSe_2$	FeSe		

Legend: insulator / semiconductor / semimetal / metal / superconductor

Figure 1.
The gallery of 2D materials.

Whereas these materials are marvelous per se, the more astounding discovery is that these 2D crystals can be combined freely to create layered compounds, paving a way for design of new functional materials and nano-devices [9, 10]. Such designer materials are called van der Waals heterostructures (vdWHs) since the atomically thin layers are not attached through a chemical reaction but rather held together via a weak van der Waals interaction. By stacking together any number of atomically thin layers, the concept provides a huge potential to tailor the unique 2D electronic states with atomic scale precision, opening the door to broaden the versatility of 2D materials and devices. Such stacked vdWHs are quite distinctive from the traditional 3D semiconductor heterostructures, as each layer acts simultaneously as the bulk material and the interface, reducing the amount of charge displacement within each layer. These vdWHs have already gained an insight into the discovery of considerably engaging physical phenomena. For instance, by combining semiconducting monolayers with graphene, one can fabricate optically active heterostructures used for photovoltaic and light-emitting devices [11–13].

Because of the charge confinement and reduced dielectric screening, the optical properties of semiconducting 2D materials are dominated by excitonic effects [14–20]. When a material goes from bulk to 2D, there is less material to screen the electric field, giving rise to an increase in Coulomb interaction and more strongly-bound electron–hole pairs (excitons). In addition, since the excitons are confined in a plane that is thinner than their Bohr radius in most 2D semiconductors, quantum confinement enhances the exciton binding energy, altering the wavelength of light they absorb and emit. These two distinctively physical phenomena naturally make the excitons bound even at room temperature with a binding energy of hundreds of meV [21]. As a consequence, such materials' two-dimensionality makes the excitons easily tunable, with a variety of external stimuli or internal stacking layers, enabling them potential candidates for various applications in optics and optoelectronics.

In this chapter, we provide a topical summary towards recent frontier research progress related to excitons in atomically thin 2D materials and vdWHs. To begin with, we clarify the different types of excitons in 2D materials, including bright and dark excitons, trions, biexcitons, and interlayer excitons. Moreover, we analyze the electronic structures and excitonic effects for two typical 2D materials (i.e., TMDs and phosphorene), as well as the excited-state dynamics in vdWHs. Furthermore, we address how external stimuli, such applied electric fields, strain, magnetic fields, and light, modulate the excitonic behavior and emission in 2D materials. Afterward, we introduce several representative optoelectronic and photonic applications based on excitonic effects of 2D materials. Finally, we give our personal insights into the challenges and outlooks in this field.

2. Exciton physics in 2D semiconductors and insulators

When the dimension of crystals converts from 3D to 2D, the electronic Coulomb screening is dramatically reduced out of quantum confinement. As a consequence, dielectric constant ϵ can fall to $\epsilon = 1$ from $\epsilon \gg 1$ in conventional bulk materials [22, 23]. Generally, the binding energies of the strongly bound excitons can reach up to 30% of the quasiparticle (QP) band gap because of the tremendous decrease in dielectric constant, rising to the magnitude of 0.1–1 eV [21, 24]. The large binding energies, which lead to a strong absorption of excitons linking to light, can not only contribute to a substantial modification in the optical spectrum both below and above the QP band gap, but also ensure a long lifetime of excitons in room temperature. Since the large binding energies of excitons in 2D monolayer hBN was initially predicted theoretically in 2006 [25], the research relating to excitons of 2D materials boomed, ranging from monolayer 2D semiconductors and insulators to vdWHs.

2.1 Excitons, trions, biexcitons, and interlayer excitons

Excitons are hydrogen-like bound states of a negatively charged electron and a positively charged hole which are attracted to each other by the electrostatic Coulomb force [26]. It is an electrically neutral quasiparticle that exists mostly in semiconductors, as well as some insulators and liquids, derived from the photo-excitation. Excitons are the main mechanism for light emission and recombination because of their large oscillator strength and enhanced light-matter interaction [27]. When it comes to low-dimension crystals, the types of excitons experience a boom. Weak dielectric screening and strong geometrical confinement mutually contribute to an extremely strong Coulomb interaction, bringing in engaging many-particle phenomena: bright and dark excitons, trions, biexcitons, and interlayer excitons.

Excitons can be bright or dark subject to the spin orientation of the individual carriers: the electron and the hole, as shown in **Figure 2(b)**. If the electron and hole have opposite spins, the two particles can easily recombine through the emission of a photon. These electron–hole pairs are called bright excitons. Whereas if they have the same spins, the electron and hole cannot easily recombine via direct emission of a photon due to the lack of required spin momentum conservation. These electron–hole pairs are called dark excitons. This darkness makes dark excitons becoming promising qubits because dark excitons cannot emit light and are thus unable to relax to a lower energy level. As a consequence, dark excitons have relatively long radiative lifetimes, lasting for over a microsecond, a period that is a thousand times longer than bright excitons and long enough to function as a qubit. By harnessing the recombination time to create 'fast' or 'slow' light, the highly stable, non-radiative nature of dark excitons paves a way for optically controlled control information

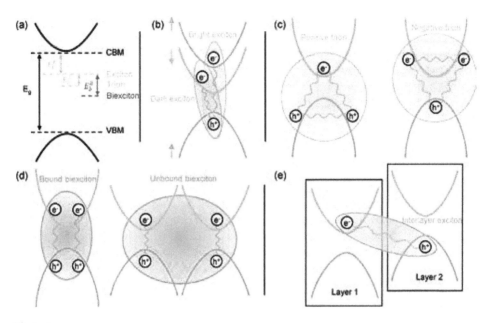

Figure 2.
Different exciton types in atomically thin nanomaterials and related heterostructures. (a) The schematic for the energy level. (b) Excitons are coulomb-bound electron hole pairs (ovals in the picture): Bright excitons consist of electrons and holes with antiparallel spins, while dark excitons consist of electrons and holes with parallel spins. (c) Trions emerge when an additional electron (hole) joins the exciton. (d) Biexcitons are created from two free excitons with different total momenta. (e) Interlayer excitons appear when electrons and holes are located in different layers.

processing. For instance, according to inducing light emission from dark excitons in monolayer WSe$_2$, it is possible to selectively control spin and valley, making dark excitons possible to encode and transport information on a chip [28, 29].

Because of the significant Coulomb interactions in 2D materials, exciton can capture an additional charge to form charged exciton known as trion, a localized excitation consisting of three charged quasiparticles (**Figure 2(c)**). Compared to exciton, a neutral electron–hole pair, trion can be negative or positive depend-ing on its charged state: a negative trion (negative e–e-h) is a complex of two electrons and one hole and a positive trion (negative e–e-h) is a complex of two holes and one electron. Trion states were predicted theoretically [30] and then observed experimentally in various 2D materials, by means of temperature-dependent photoluminescence (PL) [31] and nonlinear optical spectroscopy [32], and scanning tunneling spectroscopy [33]. Trions play a significant role in in manipulating electron spins and the valley degree of freedom for the reasons below. First, the trion binding energies are surprisingly large, reaching to about 15–45 meV in monolayer TMDs [34–36] and 100 meV in monolayer phosphorene on SiO2/Si substrate [37]. In addition, trions possess an extended population relaxation time up to tens of picoseconds [38, 39]. Finally, trions have an impact on both transport and optical properties and can be easily detected and tuned experimentally [40]. As a consequence, the electrical manipulation and detec-tion of trion, as well as its enhanced stability, make it promising for trion-based optoelectronics.

Biexcitons, also known as exciton molecules, are created from two free excitons. Biexciton configurations can be distinguished from unbound or bound biexciton cases (**Figure 2(d)**). The bound biexciton is considered as a single particle since Coulomb interaction is dominant in this complex; while the unbound biexciton is regarded as two-exciton isolated from each other because of the predominance of the repulsive Coulomb interaction [41, 42]. Similar to trions stably existing in 2D materials, biexcitons can also exist in room temperature. Among 2D materials, biexcitons were firstly observed in monolayer TMDs [43, 44], followed by predicting their binding energies of biexcitons via compu-tational simulation [45, 46].

In addition to above-mentioned intralayer excitons, interlayer excitons, where the involved electrons and holes are located in different layers, can also form in bilayer or few-layer 2D materials especially in vdWHs because of the strong Coulomb interaction (**Figure 2(e)**). After optically exciting a coherent intralayer exciton, the hole can tunnel to the other layer forming an incoherent exciton with the assistance of emission and absorption of phonons. Generally, these interlayer excitons occupy the energetically lower excitonic state than the excitons confined within one layer owing to an offset in the alignment of the monolayer band structures [47, 48]. Similar to excitons in one layer, interlayer excitons can also be either bright or dark depending on spin and momentum of the states involved [49, 50].

2.2 Excitons in atomically thin 2D materials

Among 2D semiconductors and insulators, TMDs and phosphorene have drawn tremendous attention owing to their intrinsic band gaps and strong excitonic emissions, making them potential candidates for high-performance optoelectronic applications in the visible to near-infrared regime [51]. The electronic and optical properties of 2D materials rely on their electronic band structure, which demonstrates the movement of electrons in the material and results from the periodicity of its crystal structure. When the dimension of a material degrades from bulk to 2D, the periodicity will disappear in the direction perpendicular to the plane, changing the band structure dramatically. This means by changing the number of layers in the 2D material, one can tune the band structures (e.g., a MoS_2 will become emissive when reducing to monolayer), as well as tailor the binding energies of excitons (e.g., a monolayer 2D material will absorb/emit higher energy light than a bilayer).

All TMDs have a hexagonal structure, with each monolayer consisting of the metal layer sandwiched between two chalcogenide layers (X-M-X). The two most common crystal structures are the semiconducting 2H-phase with trigonal symmetry (e.g., MoS_2, WS_2, $MoSe_2$, WSe_2, as shown in **Figure 3(a)**) and the metallic 1 T phase (e.g., WTe_2). For the semiconducting 2H-phase TMDs, they are well-known to possess an indirect band gap in bulk crystals; however, when mechanically exfoliated to a monolayer, these crystals experience a crossover from indirect to direct bandgap since the lack of interlayer interaction (**Figure 3(b)**). In addition, a decreasing layer numbers in TMDs attributes to larger absorption energy and strong photoluminescence (PL) emission in the visible spectrum, accompanying with enhanced excitonic effects, because of the reduced electronic Coulomb screening (**Figure 4(a)**) [53].

More importantly, TMDs are time-reversal symmetry but spatial inversion asymmetric. Since the strong spin–orbit coupling, the time-reversal symmetry dictates the spin splitting to have opposite spins at the K and K′ valleys of the Brillouin zones, making the excitons in TMDs are called valley excitons, which is different from the transition at the Γ valley in other 2D semiconductors such as phosphorene. As shown in **Figure 5**, the spin splitting is pretty strong in the valence band, in which spin splitting values are calculated theoretically up to 0.15 eV in 2H-MoS_2 monolayer and 0.46 eV in 2H-WSe_2 monolayer [56]. On the other hand, the broken inversion symmetry of TMD systems gives rise to a valley-dependent optical selection rule. This unique characteristic arouses the potential to control valley polarization and electronic valley. In this sense, a valley refers to the region in an electronic

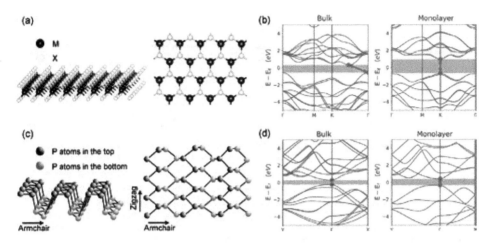

Figure 3.
Atomic structures and electronic structures of TMDs and phosphorene: Side view (left) and top view (right) of the atomic structures of the monolayer semiconducting 2H-phase TMDs (a) and of the monolayer phosphorene (c); band structures of bulk and monolayer MoS$_2$ (b) and phosphorene (d) [52]. Note that the bandgap shows a widening in phosphorene and both a widening and a crossover from indirect to direct bandgap in MoS$_2$.

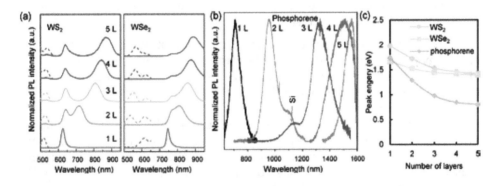

Figure 4.
The effects of layer number on the PL spectra and peak energy of TMDs and phosphorene. (a, b) normalized PL spectra of 2H-WS$_2$, 2H-WSe$_2$ and phosphorene flakes consisting of 1–5 layers. Each PL spectra is normalized to its peak intensity and system background [37, 53]. (c) Evolution of PL peak energy with layer number of 2H-WS$_2$, 2H-WSe$_2$, and phosphorene from (a, b), showing an increase in peak energy as the layer number reduces.

band structure where excitons are localized; valley polarization refers to the ratio of valley populations; and electronic valley refers a degree of freedom that is akin to charge and spin. As a consequence, optical transitions such as excitons in opposite valleys are able to be excited selectively using light with disparate chirality, paving the way to enable valleytronic devices based on photon polarizations [54, 55].

As shown in **Figure 3(c)**, phosphorene possesses a puckered orthorhombic lattice structure with P atoms distributed on two parallel planes and each P atom is covalently bonded to three adjacent atoms, resulting in strong in-plane anisotropy. Unlike TMDs that exhibit an indirect-to-direct bandgap transition when scaled down from bilayer to monolayer, phosphorene retains a direct band gap all the time, as shown in **Figure 3(d)** [57, 58]. As the layer number decrease from 5 to 1, bandgap energy of phosphorene rises remarkably because of the weaker coupling of the conduction band and the valence band caused by reduced interactions in thinner layers, showing a layer-dependent direct bandgap energies (**Figure 4(c)**). In contrast to TMDs whose PL emission occurs in the visible spectrum, the light emission

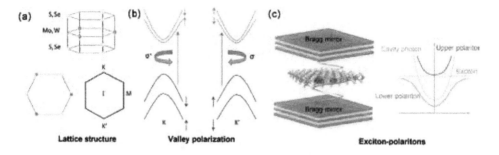

Figure 5.
Lattice structure, valley polarization, and exciton-polaritons in 2D TMDs. (a) The honeycomb lattice structure of monolayer TMDs, with broken inversion symmetry and the high-symmetry points in the first Brillouin zone. (b) Electronic bands around the K and K' points, which are spin-split by the spin–orbit interactions. The spin (up and down arrows) and valley (K and K') degrees of freedom are locked together. (c) Exciton–polariton states in a 2D semiconductor embedded inside a photonic microcavity.

of phosphorene mainly covers the near-infrared spectral regime (**Figure 4(a, b)**). Moreover, its structural anisotropy also strongly affects the excitonic effects and in phosphorene. The results from first-principles simulations demonstrate that excitonic effects can only be observed when the incident light is polarized along the armchair direction of the crystal [59].

To have an impact on excitonic effects and relevant applications, the binding energy of these quasiparticles must be clarified. As schematically illustrated in **Figure 2(a)**, the exciton binding energy is the energy difference between the electronic bandgap (E_g) and optical bandgap (E_{opt}). When higher-order excitonic quasiparticles form, more energy, i.e., the binding energy of trion or biexciton, is needed. Thus, the binding energies of exciton, trion and biexciton can be expressed as $E_b^E = E_g - E_E$, $E_b^T = E_g - E_T$, and $E_b^B = E_g - E_B$, respectively, where E_E, E_T, and E_B are emission energies of exciton, trion, and biexciton. For the most 2D conductors and insulators, a robust linear scaling law exists between the quasiparticle bandgap (E_g) and the exciton binding energy (E_b^E), namely, $E_b^E \approx E_g / 4$, regardless of their lattice configuration, bonding characteristic, and the topological property (**Figure 6**) [21]. It is worth emphasizing that the results from simulations and experiments cover almost all kinds of popular 2D monolayer semiconductors and insulators, includ-ing topological crystalline insulator (TCI) and topological insulator (TI) [60–62], TMDs [21, 63–66], nitrides (MXenes) [67], phosphorene [21, 68], IV/III–V com-pounds [21], and graphene derivatives [21]. Such an agreement between simulation and experiment results indicates that the linear scaling law can be used effectively to predict the exciton binding energy for all the 2D monolayer semiconductors and insulators. On the other hand, although comparatively lower than exciton binding energies, the binding energies of trion and biexciton in 2D materials is significantly larger than that in quasi-2D quantum wells (1–5 meV) [69]. For example, the binding energies of trion and biexciton in TMDs reach up to 45 meV and 60 meV, respectively [42, 44, 70].

2.3 Excitons in vdWHs

Composed of stacks of atomically thin 2D materials, the properties of vdWHs are determined not only by the constituent monolayers but also by the layer interac-tions. In particular, the excited-state dynamics is unique, such as the formation of interlayer excitons [47], ultrafast charge transfer between the layers [71, 72], the existence of long-lived spin and valley polarization in resident carriers [73, 74], and

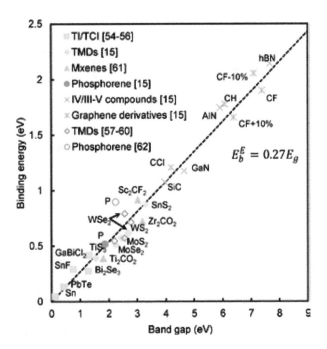

Figure 6.
Linear relationship between quasiparticle bandgap (Eg) and exciton binding energy (E b E).

moiré-trapped valley excitons in moire superlattices in vdWHs [75–78]. In terms of 2D vdWHs, the semiconducting vdWHs composed of stacked TMDC layers are the most widely studied due to their prominent exciton states and accessibility to the valley degree of freedom. More interestingly, the introduction of moiré super-lattices (**Figure 7(a)**), a periodic pattern formed by stacking two monolayer 2D materials with lattice mismatch or rotational misalignment, enables to modulate the electronic band structure and the optical properties of vdWHs [79].

After demonstrating the appearance of interlayer excitons in PL spectra, the research on exciton dynamics in vdWHs flourishes. The discovery of intralayer excitons in 2D materials can be traced back to 2015, when long-lived interlayer excitons were demonstrated in monolayer $MoSe_2/Wse_2$ heterostructures, where a pronounced additional resonance was observed at an energy below the intralayer excitons [80]. Compared with the intralayer excitons in the weak excitation regime, the PL intensity of this low-energy peak is rather prominent, which attributes to the presence of interlayer excitons as their spectral position is highly occupied. Furthermore, measuring the binding energy of interlayer excitons directly is also demonstrated in WSe_2/WS_2 heterobilayers, where a novel $1s$–$2p$ resonance are mea-sured by phase-locked mid-infrared pulses [81]. For other excited-state dynamics, such as ultrafast kinetics, long lifetimes, and moiré excitons, some research indicate they have something to do with interlayer excitons [71–78].

Empirically, charge transfer between layers of vertically stacking vdWHs is supposed to be much slow. However, transient absorption measurements, which are implemented by resonantly injecting excitons using ultrafast laser pulse, show a sub-picosecond charge separation in vdWHs: the holes injected in MoS_2 takes 200 fs transferring to $MoSe_2$ and even only 50 fs transferring to WS_2, as shown in **Figure 7(b)** [73, 74]. It is noteworthy that this process is reversible, i.e., holes trans-fer to $MoSe_2$ on the same ultrafast time scale when excitons are selectively injected in MoS_2 using excitation resonant with the higher-energy exciton feature in MoS_2.

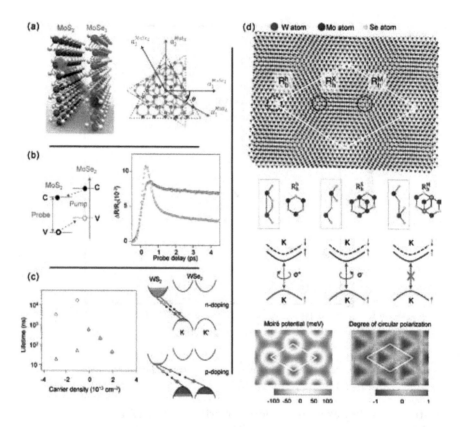

Figure 7.
Excitonic effects in vdWHs. (a) Sketch of MoS₂/MoSe₂ heterobilayer (left) and its moiré superlattice (right)
[10]. (b) Schematic of a pump-probe configuration (left), and time-resolved differential reflection of a
MoS₂/MoSe₂ heterobilayer (blue) and of MoS₂ monolayer (purple) (right) [71]. (c) Comparison between
spin-valley lifetime (circles) and hole population lifetime (triangles) under different carrier concentration in
MoS₂/MoSe₂ heterostructure (left), and schematic illustration of the interlayer electron–hole recombination
process in electron-doped and hole-doped heterostructures [74]. (d) Moiré superlattice modulates the electronic
and optical properties in WSe₂/MoSe₂ heterostructure: Three different local atomic alignments and their
corresponding schematic (top), the moiré potential of the interlayer exciton transition (left lower), and spatial
map of the optical selection rules for K-valley excitons (right lower) [76].

In addition, another interesting phenomenon is that when mismatching the bilayer vdWHs with different twist angle, the charge transfer signal keeps a constant period within 40 fs, while the recombination lifetime of these indirect excitons varies with the twist angle without any clear trend [82].

In contrast to the ultrafast charge transfer dynamics in vdWHs (<1 ps), spin and valley relaxation dynamics take place on considerably longer timescale [73, 74]. For the two distinctive relaxation processes in vdWHs (i.e., the population decay of optically excited excitons, and the exciton spin–valley lifetime which determines the information storage time in the spin– valley degree of freedom), they both are significantly longer than the monolayer case. For instance, by tuning the carrier concentration, holes' spin–valley lifetime and population lifetime possess a doping-dependent pattern in a WSe₂/WS₂ heterostructure [74]: in charge-neutral and electron-doped heterostructures (i.e., neutral and positive carrier concentrations), the spin–valley lifetime is closed to the population lifetime; nevertheless, in hole-doping heterostructures (i.e., negative carrier concentration), the spin–valley lifetime becomes orders of magnitude longer than the population lifetime (**Figure 7(c)**). The remarkable dynamics of doping-dependent lifetime attributes to the distinctive interlayer electron–hole recombi-nation process in the heterostructure, as shown in **Figure 7(c)**. In electron-doped or charge-neutral heterostructures, all holes in WSe$_2$ are pump-generated excess holes; hence, when the hole population decays to zero out of interlayer elec-tron–hole recombination, no holes can remain, let alone valley-polarized holes. The valley lifetime is thus limited by the lifetime of the total hole excess. On the contrary, in hole-doped case, the original hole density is much higher than the photo-generated density, give an equal probability for the recombination of excess electrons in WS$_2$ with holes from both valleys of WSe$_2$.

Since 2019, important breakthroughs about excitons in vdWHs has been obtained, especially three independent research simultaneously reporting the observation of moiré excitons in TMDs vdWHs, which lays a firm foundation to the engineering artificial excitonic crystals using vdWHs for nanophotonics and quantum information applications [75–77]. For example, in MoSe$_2$/WSe$_2$ hetero-bilayers with a small twist angle of ~1°, there are three points at which the local atomic registration preserves the threefold rotational symmetry Ĉ3 in the moiré supercell. The local energy extrema in the three high-symmetry points not only localizes the excitons but also provides an array of identical quantum-dot potentials (**Figure 7(d)**) [75]. The research on moiré excitons in TMDs vdWHs has been promoted after experimentally confirming the hybridization of excitonic bands that can result in a resonant enhancement of moiré superlattice effects.

3. Tuning methods of excitons

To have an impact on industrial applications especially photovoltaics, the binding energies of excitons in 2D semiconductors and insulators must be delicately designed and tuned. More importantly, these common control measures, from electrical to optical methods, function more potently in 2D materials than in 3D materials.

3.1 Electrical tuning

Since the electric field can hardly modulate the dielectric constant in monolayer 2D materials [83], early electrical tuning for excitonic behavior is mostly based on carrier density-dependent many-body Coulomb interactions, namely charged excitons or trions [84, 85]. By increasing electron doping density using different gate voltage (−100 to +80 V) in monolayer MoS$_2$ field-effect transistors, Mak et al. firstly reported the observation of tightly bound negative trions by means of absorption and photoluminescence spectroscopy [84]. These nega-tive trions hold a large trion binding energy up to ~20 meV, and can be optically created with valley and spin polarized holes. At the same time, Ross et al. also observed positive and negative trions along with neutral excitons in monolayer MoSe$_2$ field-effect transistors via photoluminescence [85]. The exciton charging effects showed a reversible electrostatic tunability, as shown in **Figure 8(a–c)**. More interestingly, the positive and negative trions exhibited a nearly identical binding energy (~30 meV), implying the same effective mass for electrons and holes. Another work demonstrated continuous tuning of the exciton binding energy in monolayer WS$_2$ field-effect transistors, finding the ground and excited excitonic states as a function of gate voltage [87].

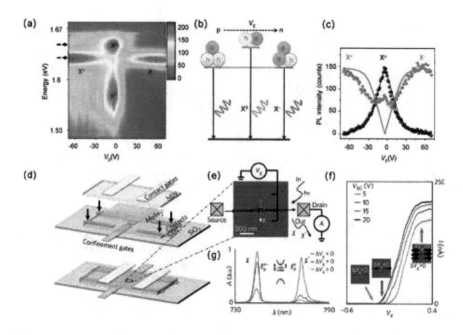

Figure 8.
Electrical tuning of excitons. (a–c) Electrical control in monolayer 2D materials [85]: (a) MoSe$_2$ PL is plotted as a function of back-gate voltage, showing a transition from positive Trion to negative Trion as gate voltage increases. (b) Illustration of the gate-dependent transitions and quasiparticles. (c) the relationship between Trion and exciton peak intensity and gate voltage at dashed arrows in (a). Solid lines are fits based on the mass action model. (d–g) Electrical control in vdWHs [86]: (d) optoelectronic transport device consisting of hBN/ MoSe$_2$/hBN heterostructure. (e) SEM image of a gate-defined monolayer MoSe$_2$ quantum dot. (f) Typically measured current across the device as a function of local gate voltage V$_g$ at different silicon backgate voltage VBG. (g) Recombination emission signals of excitons and trions as a function of emission wavelength at different V$_g$ values.

The above-mentioned works are related to monolayer 2D materials, while when it comes to heterostructures, the electrical tuning functions more efficiently [86, 88]. Employing a van der Waals heterostructure consisting of hBN/MoSe$_2$/hBN (**Figure 8(d, e)**), Wang el al. obtained homogeneous 2D electron gases by controlling disorder in TMDs, which allows for excellent electrical control of both charge and excitonic degrees of freedom [86]. Measuring the optoelectronic transport in the gate-defined heterostructure, they demonstrated gate-defined and tunable confinement of charged exciton, i.e., confinement happens when local gate voltages ΔV_g is zero or negative while being absent when ΔV_g local gate voltages are positive (**Figure 8(f)**). To further demonstrate controlled localization of charged excitons, they excited the device with a laser source at $\lambda = 660$ nm, observing both the exciton and trion recombination in PL spectra (**Figure 8(g)**). The ratio between trion and exciton recombination emission declines as ΔV_g becomes more negative, because of local depletion of trions as the device transits from the accumulation regime ($\Delta V_g > 0$) to confinement ($\Delta V_g = 0$) and depletion regimes ($\Delta V_g < 0$), respectively.

3.2 Magnetic tuning

TMDs have drawn more attention with respect to magnetic tuning than other 2D materials, since they preserve time-reversal symmetry with excitons formed at K and K′ points at the boundary of the Brillouin zone, which restricts valley polarization. However, when imposing magnetic fields, time-reversal symmetry can be broken, which splits the degeneracy between the nominally time-reversed pairs of exciton optical transitions at K and K′ valley: this is the valley Zeeman effect, as shown in **Figure 9(a, b)** [89, 91–94]. Based on the Zeeman effect, magnetic manipulation is effectively used on valley pseudospin [91], valley splitting and polarization [92], and valley angular momentums [89]. For high-order excitonic quasiparticles, valley Zeeman effect also exhibit significant effects on trions [94] and biexcitons [90] under applied magnetic fields.

In addition, magnetic fields, which change the surrounding dielectric environment, can also have an impact on the size and binding energy of excitons. By encapsulating the flakes with different materials on a monolayer WSe$_2$, Stier et al. changed the average dielectric constant, $k = (\varepsilon_t + \varepsilon_b)/2$, ranging from 1.55 to 3.0 (**Figure 9(c)**) [95]. The average energy of the field-split exciton transitions was measured in pulsed magnetic fields to 65 T, exhibiting an increasing trend with field which reveals the diamagnetic shift can infer both exciton binding energy and radius. They demonstrated increased environmental screening will enlarge exciton size but reduce exciton binding energy in 2D semiconductors, which shows a quantitatively agreement with theoretical models (**Figure 9(d)**).

3.3 Optical tuning

To control excitonic effects by breaking time-reversal symmetry in TMDs, imposing an intense circularly polarized light can also achieve the aim based on optical Stark effect, a phenomenon that photon-dressed states (Floquet states) can hybridize with the equilibrium states resulting in energy repulsion between the two states [96, 97],

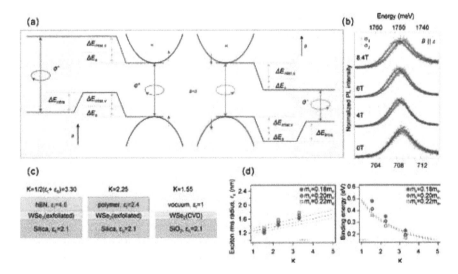

Figure 9.
Magnetic tuning of excitons. (a, b) valley Zeeman effect [89]. (a) valley Zeeman effect in a finite out-of-plane B, the degeneracy between the ±K valleys is attributed to three factors: The spin-Zeeman effect (ΔEs), the intercellular orbital magnetic moment (ΔEinter), and the intracellular contribution from the d ± id orbitals of the valence band (ΔEintra). The signs of these contributions are opposite in the two valleys. (b) Normalized polarization-resolved PL spectra of the neutral exciton peak as a function of the out-of-plane magnetic field (B), indicating a B-dependent splitting phenomenon via valley Zeeman effect. (c, d) electrical control by surrounding dielectric environment [90]: (c) the surrounding dielectric environments are changing by encapsulating hBN, polymer, or nothing on WSe$_2$ monolayer on silica substrate, where the average dielectric constant is defined as k = (εt + εb)/2 (εt and εb are the relative dielectric constants of the bottom substrate and the top encapsulation overlayer, respectively). (d) Exciton root-mean-square (rms) radius rX and exciton binding energy as a function of k (points and lines are the results from experiments and screened Keldysh model, respectively), where me, mr, and r0 are the exciton mass, the reduced mass of the exciton, and the characteristic screening length, respectively.

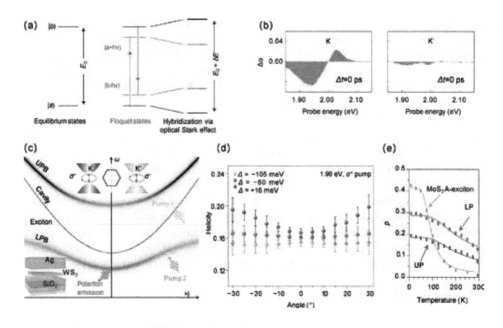

Figure 10.

Optical tuning of excitons. (a, b) optical stark effect [96]. (a) Illustration of optical stark effect for two-level system. Ground state |a⟩ and excited state |b⟩ can hybridize with Floquet states |a + ℏω⟩ and |b + ℏω⟩, bringing in shifted energy levels. (b) the valley selectivity of the optical stark effect, showing an effect only at K valley by σ − polarization pump pulses. (c–e) valley polaritons via optical pumping [98, 99]. (c) Schematic of the valley polariton phenomena. The lower polariton branch (LBP) and the upper polariton branch (UPB) are the solid curves. The valley-polarization phenomena, caused by the broken inversion symmetry, is inserted in the top. (d) Polariton emission with angle-dependent helicity. Angle-resolved helicity was measured for three detuned cavities Δ at the σ⁺ excitation, where only the positive detuned cavities shows increasing helicity as a function of angle. (e) Exciton-polaritons with a temperature-dependent emission polarization. Emission polarization for bare exciton, and upper polariton (UP) and lower polariton (LP) branches change with temperature.

as shown in **Figure 10(a, b)**. The interaction between Floquet and equilibrium states can not only bring in a wider energy level separation, but also enhance the magnitude of the energy repulsion if they are energetically close. Based on the optical Stark effect triggered off by circularly polarized light, two independent works demonstrated that the exciton level in K and K′ valleys can be selectively tuned by as much as 18 meV in WS_2 monolayer and 10 meV in WSe_2 monolayer, respectively.

Besides, optical control and manipulation have been shown effective towards valley polaritons, a half-light half-matter quasiparticles arising from hybridiza-tion of an exciton mode and a cavity mode. Owing to the large exciton binding energy and oscillator strength in TMDs, spin–valley coupling can persist at room temperature when excitons are coherently coupled to cavity photons, leading to a stable exciton-polariton formation [98–101]. Exciton polaritons are interacting bosons with very light mass, and can be independently combined in the intracav-ity and extracavity field. A schematic of the valley-polariton phenomena is shown in **Figure 10(c)**, where the microcavity structure consists of silver mirrors with a silicon dioxide cavity layer embedded with the WS_2 monolayer. The valley-polarized exciton–polaritons are optical pumped using two pumps to excite the exciton reser-voir and the lower polariton branch, showing an angle-dependent helicity because of the excitonic component of the polariton states [98]. In addition, another work based on similar method demonstrates that exciton-polaritons possess a temper-ature-dependent emission polarization, exhibiting stronger valley polarization at room temperature compared with bare excitons [99], as shown in **Figure 10(d)**.

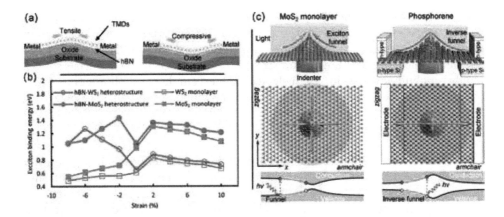

Figure 11.
Mechanical tuning of excitons. (a, b) Exciton binding energies under compressive and tensile strain [103].
(a) Schematics of hBN-TMDs heterostructures nanodevices with tensile and compressive strain. (b) Exciton
binding energies of TMD monolayer and hBN-TMD heterostructures as functions of strain. (c) Funnel effect
of Excitons under indentation. When an indenter creates an inhomogeneous strain profile that modulates the
gap, excitons (in green) in MoS_2 concentrate on isotropically the center, while excitons in phosphorene disperse,
especially along the armchair direction [109].

3.4 Mechanical tuning

2D materials possess excellent mechanical flexibility, making them stable under high compressive, tensile, and bending strain [102]. Applying mechanical strain on 2D materials, their band gaps will reduce, increase, or transit from direct to indirect, thus resulting in a strain-dependent exciton binding energy [103–108]. Based on density functional theory, Su et al. investigated the natural physical properties of TMD monolayers and hBN- TMD heterostructures, finding that they have distinctive bandgap and exciton binding energy under compressive strain (**Figure 11(a)**) [103]. MS_2 monolayers exhibit direct-to-indirect transition, while hBN-TMD heterostructures keep direct band-gap characters because of the strong charge transfer between hBN and TMD monolayers. With increasing compressive strain, the exciton binding energies of TMD monolayers gradually reduce, but the binding energies of hBN- TMD heterostructures experience a dramatically growth before decreasing (**Figure 11(b)**).

Another mechanical tuning method is by implying heterogeneous strain on 2D materials, which would result in a spatially varying bandgap with tunable exciton binding energy distribution, namely funnel effect [110–112]. In a TMDs monolayer, excitons will move towards high tensile strain region, resulting in a funnel-like band energy profile. In contrast, excitons in phosphorene are pushed away from high tensile strain region, exhibits inverse funnel effect of excitons, which is more-over highly anisotropic with more excitons flowing along the armchair direction (**Figure 11(c)**) [109]. Funnel effect is a rare method for control exciton movement, paving a way for creating a continuously varying bandgap profile in an initially homogeneous, atomically thin 2D materials.

4. Optoelectronic devices

2D materials possess strong light-matter coupling and direct band gaps from visible to infrared spectral regimes with strong excitonic resonances and large optical oscillation strength. Recent observation of valley polarization, exciton–polaritons, optically pumped lasing, exciton–polaritons, and single-photon emission highlights the potential for 2D materials for applications in novel optoelectronic devices. Combined with external stimuli, like electrical and magnetic fields, optical pumps, and strain, exciton effects in 2D materials show a highly tunability and flexibility in electroluminescent devices, photovoltaic solar cells, and photodetectors.

4.1 Electroluminescent devices

Excitonic electroluminescence (EL) emission in 2D materials is key to fully exploiting the EL devices [54, 113]. Based on different carrier injection and transport mechanisms, light-emitting devices have distinctive structures, depending on their mechanism of exciton generation: bipolar carrier injection in p-n heterojunction [114, 115], quantum well heterostructures [116], unipolar injection [117], impact excitation [118], thermal excitation [119], and interlayer excitons [48] (**Figure 12**).

As exciton emission induced by bipolar carrier injection, p-n heterojunction is the simplest device to achieving EL. Depending on the contacting way the two monolayers connect, it can be vertical or lateral. Typical p-n junction is MS_2/MSe_2 heterostructure, since their counterparts lack the caliber to function effectively [120, 121]. For instance, MoS_2 and WS_2, in which sulfur vacancies act as electron donors, are often naturally n-type; while WSe_2 and $MoSe_2$ are typically ambipolar but often unconsciously p-doped by adsorbed moisture [122, 123].

Quantum well (QW) heterostructures consist of semiconductor layer sandwiched between insulator layers and metal electrodes. EL in QW heterostructures can be observed by bipolar recombination of injected electrons and holes in the semiconductor layer when applied a bias in the metal electrodes. Since the long lifetime of carriers and enhanced exciton formation in the semiconductor layer (typical one is TMDs), the emission efficiency of multiple QW devices is much higher than that of single QW devices, and can be improved by preparing alternating layers of TMDs [116, 124].

Unipolar injection happens in a metal–insulator–semiconductor (MIS) or a semiconductor–insulator–semiconductor (SIS) heterostructures when a positive bias applied to the metal and semiconductor layers. The common insulator layer is hBN since its ability to transport holes but block electrons. If the bias is increased

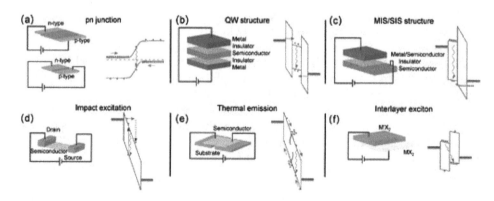

Figure 12.
EL device structures and emission mechanisms [54]. (a) Vertical and lateral p-n junctions. (b) Quantum well heterojunction structure. (c) Metal–insulator–semiconductor (MIS) and semiconductor–insulator–semiconductor (SIS) structure. (d) Lateral unipolar device where emission is induced by impact excitation. (e) Locally suspended thermal emission device. (f) Hetero-bilayer device exhibiting interlayer exciton emission. Reproduced with permission [54].

above a threshold, EL will be observed at extremely low current densities below $1\,\text{nA}\,\mu\text{m}^{-2}$, attributed to the unipolar tunneling across the hBN layer, which transfers holes from one metal/semiconductor layer (e.g., graphene and TMDs) to another electron-rich semiconductor layer (e.g., TMDs) [125].

The remaining three emission mechanism is relatively simple. For impact excitation devices, excitons are generated by impact excitation of excitons in the high field regime rather than bipolar recombination. For thermal emission devices, a semiconductor monolayer or few layers are partly suspended on a substrate, and thermal excitation and emission are evoked by locally heating the high current density regime. For bilayer emission devices, emission occurs due to the recombina-tion of electrons and holes residing in the adjacent layers.

4.2 Photovoltaic solar cells

2D materials possess large exciton binding energy with the bandgap ranging from visible to near-infrared part of the spectrum, making them attractive as candidates for photovoltaic solar cells [126, 127]. Light absorption in the active layers of a photovoltaic cell significantly determines device efficiency. To improve light absorption of 2D semiconductor photovoltaics in the ultrathin limit, light trapping designs are need, such as use of plasmonic metal particles, shells, or resonators to amplify photocurrent and photoluminescence. For large area photovoltaic applications, a common strategy is thin film interference, in which a highly reflective metal (e.g., Au or Ag) is used as a part of an "open cavity" to enhance absorption due to multipass light interactions within the semiconductor (**Figure 13(a)**) [128]. If the semiconductor layer is a monolayer absorber, an atomically thin absorber with $\lambda/4$ in thickness can be sandwiched between conductor layer and reflector layer, enabling destructive interference at the interface and thus resulting in significant absorption enhancement (**Figure 13(b)**) [129]. Another strategy to enhance light trapping is by the use of nanostructured resonators, which are coupled to or etched in thin film absorbers (**Figure 13(c, d)**) [130, 131].

Compared with free-standing monolayer with merely 10% absorption [132], the above-mentioned strategy exhibits outstanding strength for TMDC devices. For example, TMD-reflector coupled photovoltaics can have high broadband absorption of 90% and quantum efficiency of 70% [133, 134]. Accompanied with reflector, resonator, or antennas, 2D semiconductor photovoltaics trapping nearly 100% of the incident light may be achieved for nanoscale thick active layers. However, improving light absorption in sub-nanoscale thick monolay-ers faces more challenging, because not only the low absorption of monolayer (~10%) but also the limited technique to fabricate nanoscopic resonators or antennas [135].

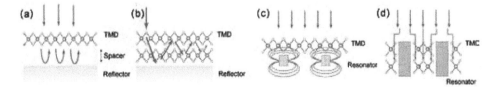

Figure 13.
Possible light trapping configurations for enhancing sunlight absorption for Photovoltaics [126]. (a) Salisbury screen-like configuration where a spacer with ~λ/4 thickness sandwichs between a low loss metal reflector and a monolayer absorber. (b) Multilayer vdWH absorber directly placed on a smooth reflective metal reflector. (c) TMD monolayer coupled with resonators/antennas. (d) Multilayer vdWH absorber etched by nanometer scale antennas/resonators. Reproduced with permission [126].

4.3 Photodetectors

Photodetection is a process converting light signals to electric signals, consisting of three physical mechanisms: light harvesting, exciton separation, and charge carrier transport to respective electrodes. According to the operation modes, photodetectors can be divided into two categories: photoconduction (i.e., photo-conductor) and photocurrent (i.e., photodiode) [136, 137]. The former one refers to the overall conductivity change out of photoexcited carriers, and the latter one involves a junction which converts photoexcited carriers into current. Generally, photoconduction-based devices possess higher quantum efficiency than photo-current-based devices, since transporting carriers can circulate many times before recombination in photoconductors. However, the response in photocurrent-based devices is faster than that in photoconduction-based devices, because of the short carrier lifetime that transporting carriers (electrons and holes) are both involved in the photocurrent generation and recombine with their counterpart after reach-ing to their own electrodes.

Two common 2D materials used for photodetectors are graphene [138–140] and TMDs (**Figure 14**) [55, 143, 144]. Based on photothermal with weak photovoltaic effect, graphene photodetectors usually show higher dark currents and smaller responsivity, but much wider operational bandwidths. In contrast, TMDs photode-tectors operating on photovoltaic effect, exhibiting lower dark currents and higher responsivity.

For graphene-based photoconductors, typical devices are hybrid, adding a light absorption material, such as quantum dots [145], perovskites [146], silicon [147], carbon nanotubes [148], and TMDs [141], as active layer to improve the responsivity. For graphene-based photodiodes, this earliest reported one is metal–graphene–metal photodiodes, in which photocurrent was generated by local illumination of the metal/graphene interfaces of a back-gated graphene field-effect transistor. The

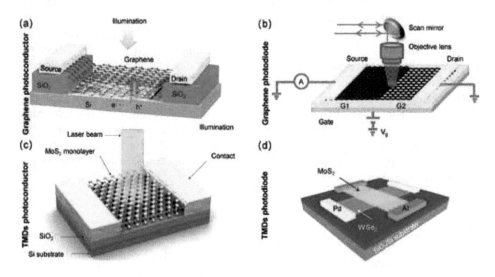

Figure 14.
Typical 2D photodetectors. (a) Schematic of a hybrid graphene photoconductor [141]. (b) Schematic of a single-bilayer graphene interface junction, in which photocurrent generation is dominant by photothermoelectric effect [142]. (c) Schematic of monolayer MoS₂ lateral photoconductor [143]. (d) Schematic of vertical p–n photodiode formed by monolayer MoS₂ and WSe₂, in which a photocurrent hot spot is produced at the heterojunction [120].

resulting current can be attributed to either photovoltaic effect [149] or photo-ther-moelectric effect [142]. To additionally improve the performance, common struc-tures are graphene–semiconductor heterojunction photodiodes, in which planar junctions of graphene and group-IV elements or other compound semiconductors act as Schottky diodes [150, 151].

For TMD photodetectors, devices can have in-plane or out-of-plane structures, based on the semiconductor layers stacking laterally or vertically. In-plane devices take advantage of better control of the material s properties via electrostatic gating [143]. But out-of-plane devices can bear a much higher bias field (up to ~ 1 V nm^{-1}), enabling a reduced excitonic binding energy in multilayer structures for more efficient exciton dissociations [152]. TMDs-based photoconductors are usually enhanced by illuminating the semiconductor–metal contacts [40] and in short-channel devices [153], and their conductance can be changed by doping and trap-ping of photogenerated carriers by impurity states [154, 155]. On the other hand, TMDs-based photodiodes exhibit higher tunability based on the photocurrent mode, consisting of an in-plane or out-of-plane junction where a built-in electric field is created [120, 156, 157]. In this situation, electrostatic gates can further tune the device doping levels, owing to the very small interlayer separation (<1 nm) which produces extremely high built-in electric fields (~ 1 V nm^{-1}).

5. Summary and perspective

In summary, 2D materials exhibit excitonic effects due to spatial confinement and reduced screening at the 2D limit, resulting in fascinating many-particle phenomena, such as excitons, trions, biexcitons, and interlayer excitons. Enhanced binding energies owing to the strong Coulomb interaction make these quasipar-ticles easy to characterize and control. In addition, the sensitivity of these quasi-particles to a variety of external stimuli allows the possibility of modulating the inherent optical, electrical, and optoelectronic properties of 2D materials, making them potential candidates for novel optoelectronic applications.

In addition to well-studied 2D materials, such as graphene, phosphorene, and TMDs, the family of 2D crystals is continuously growing, making excitonic effects versatile in different 2D systems. In particular, assembling vdWHs, which now can be mechanically assembled or grown by ample methods, can open up a new route for exploring unique exciton physics and applications. For example, an in-plane moiré superlattice, formed by vertically stacking two monolayer semiconductors mismatching or rotationally misaligning, can modulate the electronic band struc-ture and thus lead to electronic phenomena, such as fractal quantum Hall effect, unconventional superconductivity, and tunable Mott insulators.

Acknowledgements

This work is supported by the start-up funds at Stevens Institute of Technology.

Author details

Xiaoyang Zheng and Xian Zhang*
Stevens Institute of Technology, Hoboken, New Jersey, United States

*Address all correspondence to: xzhang4@stevens.edu

References

[1] Novoselov KS, Geim AK, Morozov SV, Jiang D, Zhang Y, Dubonos SV, et al. Electric field effect in atomically thin carbon films. Science. 2004;**306**(5696):666-669

[2] Jin C, Lin F, Suenaga K, Iijima S. Fabrication of a freestanding boron nitride single layer and its defect assignments. Physical Review Letters. 2009;**102**(19):195505

[3] Liu H, Neal AT, Zhu Z, Luo Z, Xu X, Tománek D, et al. Phosphorene: An unexplored 2D semiconductor with a high hole mobility. ACS Nano. 2014;**8**(4):4033-4041

[4] Molle A, Goldberger J, Houssa M, Xu Y, Zhang SC, Akinwande D. Buckled two-dimensional Xene sheets. Nature Materials. 2017;**16**(2):163

[5] Manzeli S, Ovchinnikov D, Pasquier D, Yazyev OV, Kis A. 2D transition metal dichalcogenides. Nature Reviews Materials. 2017;**2**(8):17033

[6] Mas-Balleste R, Gomez-Navarro C, Gomez-Herrero J, Zamora F. 2D materials: To graphene and beyond. Nanoscale. 2011;**3**(1):20-30

[7] Li Y, Ye F, Xu J, Zhang W, Feng P, Zhang X. Gate-tuned temperature in a hexagonal boron nitride-encapsulated 2D semiconductor devices. IEEE Transactions on Electronic Devices. 2018;**99**:1. DOI: 10.1109/TED.2018.2851945

[8] Zhang X. Characterization of layer number of two-dimensional transition metal diselenide semiconducting devices using Si-peak analysis. Advances in Materials Science and Engineering. 2019;**2019**:7865698. DOI: 10.1155/2019/7865698

[9] Geim AK, Grigorieva IV. Van der Waals heterostructures. Nature. 2013;**499**(7459):419

[10] Novoselov KS, Mishchenko A, Carvalho A, Neto AC. 2D materials and van der Waals heterostructures. Science. 2016;**353**(6298):aac9439

[11] Furchi MM, Zechmeister AA, Hoeller F, Wachter S, Pospischil A, Mueller T. Photovoltaics in Van der Waals heterostructures. IEEE Journal of Selected Topics in Quantum Electronics. 2016;**23**(1):106-116

[12] Furchi MM, Höller F, Dobusch L, Polyushkin DK, Schuler S, Mueller T. Device physics of van der Waals heterojunction solar cells. npj 2D Materials and Applications. 2018;**2**(1):3

[13] Liu CH, Clark G, Fryett T, Wu S, Zheng J, Hatami F, et al. Nanocavity integrated van der Waals heterostructure light-emitting tunneling diode. Nano Letters. 2016;**17**(1):200-205

[14] Pei J, Yang J, Yildirim T, Zhang H, Lu Y. Many-body complexes in 2D semiconductors. Advanced Materials. 2019;**31**(2):1706945

[15] Xiao J, Zhao M, Wang Y, Zhang X. Excitons in atomically thin 2D semiconductors and their applications. Nano. 2017;**6**(6):1309-1328

[16] Mueller T, Malic E. Exciton physics and device application of two-dimensional transition metal dichalcogenide semiconductors. npj 2D Materials and Applications. 2018;**2**(1):29

[17] Van der Donck M, Zarenia M, Peeters FM. Excitons and trions in monolayer transition metal dichalcogenides: A comparative study between the multiband model and the quadratic single-band model. Physical Review B. 2017;**96**:035131

[18] Van der Donck M, Zarenia M, Peeters FM. Excitons, trions, and

biexcitons in transition-metal dichalc-ogenides: Magnetic-field dependence. Physical Review B. 2018;**97**:195408

[19] Van der Donck M, Peeters FM. Interlayer excitons in transition metal dichalcogenide heterostructures. Physical Review B. 2018;**98**:115104

[20] Van der Donck M, Peeters FM. Excitonic complexes in anisotropic atomically thin two-dimensional materials: Black phosphorus and TiS_3. Physical Review B. 2018;**98**:235401

[21] Jiang Z, Liu Z, Li Y, Duan W. Scaling universality between band gap and exciton binding energy of two-dimensional semiconductors. Physical Review Letters. 2017;**118**(26):266401

[22] Laturia A, Van de Put ML, Vandenberghe WG. Dielectric pro-perties of hexagonal boron nitride and transition metal dichalcogenides: From monolayer to bulk. npj 2D Materials and Applications. 2018;**2**(1):6

[23] Hwang EH, Sarma SD. Dielectric function, screening, and plasmons in two-dimensional graphene. Physical Review B. 2007;**75**(20):205418

[24] Kidd DW, Zhang DK, Varga K. Binding energies and structures of two-dimensional excitonic complexes in transition metal dichalcogenides. Physical Review B. 2016;**93**(12):125423

[25] Wirtz L, Marini A, Rubio A. Excitons in boron nitride nanotubes: Dimensionality effects. Physical Review Letters. 2006;**96**(12):126104

[26] Elliott RJ. Intensity of optical absorption by excitons. Physical Review. 1957;**108**(6):1384

[27] Citrin DS. Radiative lifetimes of excitons in quantum wells: Localization and phase-coherence effects. Physical Review B. 1993;**47**(7):3832

[28] Zhang XX, Cao T, Lu Z, Lin YC, Zhang F, Wang Y, et al. Magnetic brightening and control of dark excitons in monolayer WSe 2. Nature Nanotechnology. 2017;**12**(9):883

[29] Zhou Y, Scuri G, Wild DS, High AA, Dibos A, Jauregui LA, et al. Probing dark excitons in atomically thin semiconductors via near-field coupling to surface plasmon polaritons. Nature Nanotechnology. 2017;**12**(9):856

[30] Ganchev B, Drummond N, Aleiner I, Fal'ko V. Three-particle complexes in two-dimensional semiconductors. Physical Review Letters. 2015;**114**(10):107401

[31] Sun H, Wang J, Wang F, Xu L, Jiang K, Shang L, et al. Enhanced exciton emission behavior and tunable band gap of ternary W (S_xSe_{1-x}) 2 monolayer: Temperature dependent optical evidence and first-principles calculations. Nanoscale. 2018;**10**(24):11553-11563

[32] Ye J, Yan T, Niu B, Li Y, Zhang X. Nonlinear dynamics of trions under strong optical excitation in monolayer MoSe 2. Scientific Reports. 2018;**8**(1):2389

[33] Demeridou I, Paradisanos I, Liu Y, Pliatsikas N, Patsalas P, Germanis S, et al. Spatially selective reversible charge carrier density tuning in WS_2 monolayers via photochlorination. 2D Materials. 2018;**6**(1):015003

[34] Zhang DK, Kidd DW, Varga K. Excited biexcitons in transition metal dichalcogenides. Nano Letters. 2015;**15**(10):7002-7005

[35] Berkelbach TC, Hybertsen MS, Reichman DR. Theory of neutral and charged excitons in monolayer transition metal dichalcogenides. Physical Review B. 2013;**88**(4):045318

[36] Cadiz F, Tricard S, Gay M, Lagarde D, Wang G, Robert C, et al.

Well separated Trion and neutral excitons on superacid treated MoS_2 monolayers. Applied Physics Letters. 2016;**108**(25):251106

[37] Yang J, Xu R, Pei J, Myint YW, Wang F, Wang Z, et al. Optical tuning of exciton and Trion emissions in monolayer phosphorene. Light: Science & Applications. 2015;**4**(7):e312

[38] Gao F, Gong Y, Titze M, Almeida R, Ajayan PM, Li H. Valley Trion dynamics in monolayer $MoSe_2$. Physical Review B. 2016;**94**(24):245413

[39] Wang G, Bouet L, Lagarde D, Vidal M, Balocchi A, Amand T, et al. Valley dynamics probed through charged and neutral exciton emission in monolayer WSe_2. Physical Review B. 2014;**90**(7):075413

[40] Mak KF, McGill KL, Park J, McEuen PL. The valley hall effect in MoS_2 transistors. Science. 2014;**344** (6191):1489-1492

[41] Sahin M, Koç F. A model for the recombination and radiative lifetime of trions and biexcitons in spherically shaped semiconductor nanocrystals. Applied Physics Letters. 2013;**102**(18):183103

[42] Steinhoff A, Florian M, Singh A, Tran K, Kolarczik M, Helmrich S, et al. Biexciton fine structure in monolayer transition metal dichalcogenides. Nature Physics. 2018;**14**(12):1199

[43] You Y, Zhang XX, Berkelbach TC, Hybertsen MS, Reichman DR, Heinz TF. Observation of biexcitons in monolayer WSe_2. Nature Physics. 2015;**11**(6):477

[44] Shang J, Shen X, Cong C, Peimyoo N, Cao B, Eginligil M, et al. Observation of excitonic fine structure in a 2D transition-metal dichalcogenide semiconductor. ACS Nano. 2015;**9**(1):647-655

[45] Mayers MZ, Berkelbach TC, Hybertsen MS, Reichman DR. Binding energies and spatial structures of small carrier complexes in monolayer transition-metal dichalcogenides via diffusion Monte Carlo. Physical Review B. 2015;**92**(16):161404

[46] Szyniszewski M, Mostaani E, Drummond ND, Fal'Ko VI. Binding energies of trions and biexcitons in two-dimensional semiconductors from diffusion quantum Monte Carlo calculations. Physical Review B. 2017;**95**(8):081301

[47] Ovesen S, Brem S, Linderälv C, Kuisma M, Korn T, Erhart P, et al. Interlayer exciton dynamics in van der Waals heterostructures. Communications on Physics. 2019;**2**(1):23

[48] Ross JS, Rivera P, Schaibley J, Lee-Wong E, Yu H, Taniguchi T, et al. Interlayer exciton optoelectronics in a 2D heterostructure p–n junction. Nano Letters. 2017;**17**(2):638-643

[49] Chen Y, Quek SY. Tunable bright interlayer excitons in few-layer black phosphorus based van der Waals heterostructures. 2D Materials. 2018;**5** (4):045031

[50] Latini S, Winther KT, Olsen T, Thygesen KS. Interlayer excitons and band alignment in MoS_2/hBN/WSe_2 van der Waals heterostructures. Nano Letters. 2017;**17**(2):938-945

[51] Yu S, Wu X, Wang Y, Guo X, Tong L. 2D materials for optical modulation: Challenges and opportunities. Advanced Materials. 2017;**29**(14):1606128

[52] Available from: https://www.ossila. com/pages/graphene-2d-materials[Acc-essed: 14 July 2019]

[53] Zhao W, Ghorannevis Z, Chu L, Toh M, Kloc C, Tan PH, et al. Evolution of electronic structure in atomically thin

sheets of WS_2 and WSe_2. ACS Nano. 2012;7(1):791-797

[54] Mak KF, Shan J. Photonics and optoelectronics of 2D semiconductor transition metal dichalcogenides. Nature Photonics. 2016;10(4):216

[55] Wang J, Verzhbitskiy I, Eda G. Electroluminescent devices based on 2D semiconducting transition metal dichalcogenides. Advanced Materials. 2018;30(47):1802687

[56] Zhu ZY, Cheng YC, Schwingenschlögl U. Giant spin-orbit-induced spin splitting in two-dimensional transition-metal dichalcogenide semiconductors. Physical Review B. 2011;84 (15):153402

[57] Zhang S, Yang J, Xu R, Wang F, Li W, Ghufran M, et al. Extraordinary photoluminescence and strong temperature/angle-dependent Raman responses in few-layer phosphorene. ACS Nano. 2014;8(9):9590-9596

[58] Qiao J, Kong X, Hu ZX, Yang F, Ji W. High-mobility transport anisotropy and linear dichroism in few-layer black phosphorus. Nature Communications. 2014;5:4475

[59] Tran V, Soklaski R, Liang Y, Yang L. Layer-controlled band gap and anisotropic excitons in few-layer black phosphorus. Physical Review B. 2014;89(23):235319

[60] Xu Y, Yan B, Zhang HJ, Wang J, Xu G, Tang P, et al. Large-gap quantum spin hall insulators in tin films. Physical Review Letters. 2013;111(13):136804

[61] Liu J, Qian X, Fu L. Crystal field effect induced topological crystalline insulators in monolayer IV–VI semiconductors. Nano Letters. 2015;15 (4):2657-2661

[62] Li L, Zhang X, Chen X, Zhao M. Giant topological nontrivial band gaps in chloridized gallium bismuthide. Nano Letters. 2015;15(2):1296-1301

[63] Ugeda MM, Bradley AJ, Shi SF, Felipe H, Zhang Y, Qiu DY, et al. Giant bandgap renormalization and excitonic effects in a monolayer transition metal dichalcogenide semiconductor. Nature Materials. 2014;13(12):1091

[64] Klots AR, Neważ AK, Wang B, Prasai D, Krzyzanowska H, Lin J, et al. Probing excitonic states in suspended two-dimensional semiconductors by photocurrent spectroscopy. Scientific Reports. 2014;4:6608

[65] Zhu B, Chen X, Cui X. Exciton binding energy of monolayer WS_2. Scientific Reports. 2015;5:9218

[66] Hanbicki AT, Currie M, Kioseoglou G, Friedman AL, Jonker BT. Measurement of high exciton binding energy in the monolayer transition-metal dichalcogenides WS_2 and WSe_2. Solid State Communications. 2015; 203:16-20

[67] Khazaei M, Arai M, Sasaki T, Chung CY, Venkataramanan NS, Estili M, et al. Novel electronic and magnetic properties of two-dimensional transition metal carbides and nitrides. Advanced Functional Materials. 2013;23(17):2185-2192

[68] Wang X, Jones AM, Seyler KL, Tran V, Jia Y, Zhao H, et al. Highly anisotropic and robust excitons in monolayer black phosphorus. Nature Nanotechnology. 2015;10(6):517

[69] Zhang X. Excitonic structure in atomically-thin transition metal dichalcogenides [doctoral dissertation]. Columbia University.

[70] Plechinger G, Nagler P, Kraus J, Paradiso N, Strunk C, Schüller C, et al. Identification of excitons, trions and biexcitons in single-layer WS_2. Physica

Status Solidi RRL: Rapid Research Letters. 2015;**9**(8):457-461

[71] Ceballos F, Bellus MZ, Chiu HY, Zhao H. Ultrafast charge separation and indirect exciton formation in a MoS_2–$MoSe_2$ van der Waals heterostructure. ACS Nano. 2014;**8**(12):12717-12724

[72] Hong X, Kim J, Shi SF, Zhang Y, Jin C, Sun Y, et al. Ultrafast charge transfer in atomically thin MoS_2/WS_2 heterostructures. Nature Nanotechnology. 2014;**9**(9):682

[73] Kim J, Jin C, Chen B, Cai H, Zhao T, Lee P, et al. Observation of ultralong valley lifetime in WSe_2/MoS_2 heterostructures. Science Advances. 2017;**3**(7):e1700518

[74] Jin C, Kim J, Utama MI, Regan EC, Kleemann H, Cai H, et al. Imaging of pure spin-valley diffusion current in WS_2-WSe_2 heterostructures. Science. 2018;**360**(6391):893-896

[75] Seyler KL, Rivera P, Yu H, Wilson NP, Ray EL, Mandrus DG, et al. Signatures of moiré-trapped valley excitons in $MoSe_2$/WSe_2 hetero-bilayers. Nature. 2019;**567** (7746):66

[76] Tran K, Moody G, Wu F, Lu X, Choi J, Kim K, et al. Evidence for moiré excitons in van der Waals heterostructures. Nature. 2019;**567** (7746):71

[77] Jin C, Regan EC, Yan A, Utama MI, Wang D, Zhao S, et al. Observation of moiré excitons in WSe_2/WS_2 heterostructure superlattices. Nature. 2019; **567**(7746):76

[78] Alexeev EM, Ruiz-Tijerina DA, Danovich M, Hamer MJ, Terry DJ, Nayak PK, et al. Resonantly hybridized excitons in moiré superlattices in van der Waals heterostructures. Nature. 2019;**567** (7746):81

[79] Zhang C, Chuu CP, Ren X, Li MY, Li LJ, Jin C, et al. Interlayer couplings, Moiré patterns, and 2D electronic superlattices in MoS_2/WSe_2 hetero-bilayers. Science Advances. 2017;**3**(1):e1601459

[80] Rivera P, Schaibley JR, Jones AM, Ross JS, Wu S, Aivazian G, et al. Observation of long-lived interlayer excitons in monolayer $MoSe_2$–WSe_2 heterostructures. Nature Communications. 2015;**6**:6242

[81] Merkl P, Mooshammer F, Steinleitner P, Girnghuber A, Lin KQ, Nagler P, et al. Ultrafast transition between exciton phases in van der Waals heterostructures. Nature Materials. 2019;**18**(7):691-696

[82] Zhu H, Wang J, Gong Z, Kim YD, Hone J, Zhu XY. Interfacial charge transfer circumventing momentum mismatch at two-dimensional van der Waals heterojunctions. Nano Letters. 2017;**17**(6):3591-3598

[83] Santos EJ, Kaxiras E. Electrically driven tuning of the dielectric constant in MoS_2 layers. ACS Nano. 2013;**7**(12):10741-10746

[84] Mak KF, He K, Lee C, Lee GH, Hone J, Heinz TF, et al. Tightly bound trions in monolayer MoS_2. Nature Materials. 2013;**12**(3):207

[85] Ross JS, Wu S, Yu H, Ghimire NJ, Jones AM, Aivazian G, et al. Electrical control of neutral and charged excitons in a monolayer semiconductor. Nature Communications. 2013;**4**:1474

[86] Wang K, De Greve K, Jauregui LA, Sushko A, High A, Zhou Y, et al. Electrical control of charged carriers and excitons in atomically thin materials. Nature Nanotechnology. 2018;**13**(2):128

[87] Chernikov A, van der Zande AM, Hill HM, Rigosi AF, Velauthapillai A, Hone J, et al. Electrical tuning of exciton binding energies in monolayer

WS$_2$. Physical Review Letters. 2015;**115**(12):126802

[88] Ciarrocchi A, Unuchek D, Avsar A, Watanabe K, Taniguchi T, Kis A. Polarization switching and electrical control of interlayer excitons in two-dimensional van der Waals heterostructures. Nature Photonics. 2019; **13**(2):131

[89] Srivastava A, Sidler M, Allain AV, Lembke DS, Kis A, Imamoğlu A. Valley Zeeman effect in elementary optical excitations of monolayer WSe$_2$. Nature Physics. 2015;**11**(2):141

[90] Stier AV, Wilson NP, Clark G, Xu X, Crooker SA. Probing the influence of dielectric environment on excitons in monolayer WSe$_2$: Insight from high magnetic fields. Nano Letters. 2016; **16**(11):7054-7060

[91] Aivazian G, Gong Z, Jones AM, Chu RL, Yan J, Mandrus DG, et al. Magnetic control of valley pseudospin in monolayer WSe$_2$. Nature Physics. 2015;**11**(2):148

[92] Li Y, Ludwig J, Low T, Chernikov A, Cui X, Arefe G, et al. Valley splitting and polarization by the Zeeman effect in monolayer MoSe$_2$. Physical Review Letters. 2014;**113**(26):266804

[93] MacNeill D, Heikes C, Mak KF, Anderson Z, Kormányos A, Zólyomi V, et al. Breaking of valley degeneracy by magnetic field in monolayer MoSe$_2$. Physical Review Letters. 2015;**114** (3):037401

[94] Lyons TP, Dufferwiel S, Brooks M, Withers F, Taniguchi T, Watanabe K, et al. The valley Zeeman effect in inter- and intra-valley trions in monolayer WSe$_2$. Nature Communications. 2019; **10**(1):2330

[95] Stevens CE, Paul J, Cox T, Sahoo PK, Gutiérrez HR, Turkowski V, et al. Biexcitons in monolayer transition metal dichalcogenides tuned by magnetic fields. Nature Communications. 2018;**9**(1):3720

[96] Sie EJ, McIver JW, Lee YH, Fu L, Kong J, Gedik N. Valley-selective optical stark effect in monolayer WS$_2$. Nature Materials. 2014;**14**:290-294

[97] Kim J, Hong X, Jin C, Shi SF, Chang CY, Chiu MH, et al. Ultrafast generation of pseudo-magnetic field for valley excitons in WSe$_2$ monolayers. Science. 2014;**346**(6214):1205-1208

[98] Sun Z, Gu J, Ghazaryan A, Shotan Z, Considine CR, Dollar M, et al. Optical control of room-temperature valley polaritons. Nature Photonics. 2017;**11**(8):491

[99] Chen YJ, Cain JD, Stanev TK, Dravid VP, Stern NP. Valley-polarized exciton–polaritons in a monolayer semiconductor. Nature Photonics. 2017;**11**(7):431

[100] Dufferwiel S, Lyons TP, Solnyshkov DD, Trichet AA, Withers F, Schwarz S, et al. Valley-addressable polaritons in atomically thin semiconductors. Nature Photonics. 2017; **11**(8):497

[101] Zeng H, Dai J, Yao W, Xiao D, Cui X. Valley polarization in MoS$_2$ monolayers by optical pumping. Nature Nanotechnology. 2012;**7**(8):490

[102] Kim SJ, Choi K, Lee B, Kim Y, Hong BH. Materials for flexible, stretchable electronics: Graphene and 2D materials. Annual Review of Materials Research. 2015;**45**:63-84

[103] Su J, He J, Zhang J, Lin Z, Chang J, Zhang J, et al. Unusual properties and potential applications of strain BN-MS$_2$ (M=Mo, W) heterostructures. Scientific Reports. 2019;**9**(1):3518

[104] Defo RK, Fang S, Shirodkar SN, Tritsaris GA, Dimoulas A, Kaxiras E.

Strain dependence of band gaps and exciton energies in pure and mixed transition-metal dichalcogenides. Physical Review B. 2016;**94**(15):155310

[105] Kumar A, Ahluwalia PK. Mechanical strain dependent electronic and dielectric properties of two-dimensional honeycomb structures of MoX_2 (X= S, Se, Te). Physica B: Condensed Matter. 2013;**419**:66-75

[106] Hu Y, Zhang F, Titze M, Deng B, Li H, Cheng GJ. Straining effects in MoS_2 monolayer on nanostructured substrates: Temperature-dependent photoluminescence and exciton dynamics. Nanoscale. 2018;**10**(12): 5717-5724

[107] Arra S, Babar R, Kabir M. Exciton in phosphorene: Strain, impurity, thickness, and heterostructure. Physical Review B. 2019;**99**(4): 045432

[108] Aslan OB, Datye IM, Mleczko MJ, Sze Cheung K, Krylyuk S, Bruma A, et al. Probing the optical properties and strain-tuning of ultrathin Mo_{1-x} W_xTe_2. Nano Letters. 2018;**18**(4):2485-2491

[109] San-Jose P, Parente V, Guinea F, Roldán R, Prada E. Inverse funnel effect of excitons in strained black phosphorus. Physical Review X. 2016;**6**(3):031046

[110] Feng J, Qian X, Huang CW, Li J. Strain-engineered artificial atom as a broad-spectrum solar energy funnel. Nature Photonics. 2012;**6**(12):866

[111] Krustok J, Kaupmees R, Jaaniso R, Kiisk V, Sildos I, Li B, et al. Local strain-induced band gap fluctuations and exciton localization in aged WS_2 monolayers. AIP Advances. 2017;7(6):065005

[112] Castellanos-Gomez A, Roldán R, Cappelluti E, Buscema M, Guinea F, van der Zant HS, et al. Local strain

engineering in atomically thin MoS_2. Nano Letters. 2013;**13**(11):5361-5366

[113] Wu S, Buckley S, Jones AM, Ross JS, Ghimire NJ, Yan J, et al. Control of two-dimensional excitonic light emission via photonic crystal. 2D Materials. 2014;**1**(1):011001

[114] Ye Y, Ye Z, Gharghi M, Zhu H, Zhao M, Wang Y, et al. Exciton-dominant electroluminescence from a diode of monolayer MoS_2. Applied Physics Letters. 2014;**104**(19):193508

[115] Cheng R, Li D, Zhou H, Wang C, Yin A, Jiang S, et al. Electroluminescence and photocurrent generation from atomically sharp WSe_2/MoS_2 heterojunction p–n diodes. Nano Letters. 2014;**14**(10): 5590-5597

[116] Withers F, Del Pozo-Zamudio O, Mishchenko A, Rooney AP, Gholinia A, Watanabe K, et al. Light-emitting diodes by band-structure engineering in van der Waals heterostructures. Nature Materials. 2015;**14**(3):301

[117] Wang S, Wang J, Zhao W, Giustiniano F, Chu L, Verzhbitskiy I, et al. Efficient carrier-to-exciton conversion in field emission tunnel diodes based on MIS-type van der Waals heterostack. Nano Letters. 2017;**17**(8):5156-5162

[118] Sundaram RS, Engel M, Lombardo A, Krupke R, Ferrari AC, Avouris P, et al. Electroluminescence in single layer MoS_2. Nano Letters. 2013;**13**(4): 1416-1421

[119] Kim YD, Gao Y, Shiue RJ, Wang L, Aslan OB, Bae MH, et al. Ultrafast graphene light emitters. Nano Letters. 2018;**18**(2):934-940

[120] Lee CH, Lee GH, Van Der Zande AM, Chen W, Li Y, Han M, et al. Atomically thin p–n junctions with van der Waals heterointerfaces. Nature Nanotechnology. 2014;**9**(9):676

[121] Furchi MM, Pospischil A, Libisch F, Burgdörfer J, Mueller T. Photovoltaic effect in an electrically tunable van der Waals heterojunction. Nano Letters. 2014;**14**(8):4785-4791

[122] Qiu H, Xu T, Wang Z, Ren W, Nan H, Ni Z, et al. Hopping transport through defect-induced localized states in molybdenum disulphide. Nature Communications. 2013;**4**:2642

[123] Schmidt H, Giustiniano F, Eda G. Electronic transport properties of transition metal dichalcogenide field-effect devices: Surface and interface effects. Chemical Society Reviews. 2015;**44**(21):7715-7736

[124] Withers F, Del Pozo-Zamudio O, Schwarz S, Dufferwiel S, Walker PM, Godde T, et al. WSe$_2$ light-emitting tunneling transistors with enhanced brightness at room temperature. Nano Letters. 2015;**15**(12):8223-8228

[125] Li D, Cheng R, Zhou H, Wang C, Yin A, Chen Y, et al. Electric-field-induced strong enhancement of electroluminescence in multilayer molybdenum disulfide. Nature Communications. 2015;**6**:7509

[126] Jariwala D, Davoyan AR, Wong J, Atwater HA. Van der Waals materials for atomically-thin photovoltaics: Promise and outlook. ACS Photonics. 2017;**4**(12):2962-2970

[127] Ganesan VD, Linghu J, Zhang C, Feng YP, Shen L. Heterostructures of phosphorene and transition metal dichalcogenides for excitonic solar cells: A first-principles study. Applied Physics Letters. 2016;**108**(12):122105

[128] Jang MS, Brar VW, Sherrott MC, Lopez JJ, Kim LK, Kim S, et a. Tunable large resonant absorption in a Mid-IR graphene salisbury screen. arXiv preprint arXiv:1312.6463; 2013

[129] Aydin K, Ferry VE, Briggs RM, Atwater HA. Broadband polarization-independent resonant light absorption using ultrathin plasmonic super absorbers. Nature Communications. 2011;**2**:517

[130] Piper JR, Fan S. Total absorption in a graphene monolayer in the optical regime by critical coupling with a photonic crystal guided resonance. ACS Photonics. 2014;**1**(4):347-353

[131] Kim SJ, Fan P, Kang JH, Brongersma ML. Creating semicond-uctor metafilms with designer absorption spectra. Nature Communications. 2015;**6**:7591

[132] Mak KF, Lee C, Hone J, Shan J, Heinz TF. Atomically thin MoS$_2$: A new direct-gap semiconductor. Physical Review Letters. 2010;**105**(13):136805

[133] Jariwala D, Davoyan AR, Tagliabue G, Sherrott MC, Wong J, Atwater HA. Near-unity absorption in van der Waals semiconductors for ultrathin optoelectronics. Nano Letters. 2016;**16**(9):5482-5487

[134] Wong J, Jariwala D, Tagliabue G, Tat K, Davoyan AR, Sherrott MC, et al. High photovoltaic quantum efficiency in ultrathin van der Waals heterostructures. ACS Nano. 2017;**11**(7):7230-7240

[135] Bahauddin SM, Robatjazi H, Thomann I. Broadband absorption engineering to enhance light absorption in monolayer MoS$_2$. ACS Photonics. 2016;**3**(5):853-862

[136] Sze SM. Semiconductor Devices: Physics and Technology. Hoboken, New Jersey, United States: John Wiley & Sons; 2008

[137] Konstantatos G, Sargent EH. Nanostructured materials for photon detection. Nature Nanotechnology. 2010;**5**(6):391

[138] Koppens FH, Mueller T, Avouris P, Ferrari AC, Vitiello MS,

Polini M. Photodetectors based on graphene, other two-dimensional materials and hybrid systems. Nature Nanotechnology. 2014;**9**(10):780

[139] Sun Z, Chang H. Graphene and graphene-like two-dimensional materials in photodetection: Mechanisms and methodology. ACS Nano. 2014; **8**(5):4133-4156

[140] Xia F, Mueller T, Lin YM, Valdes-Garcia A, Avouris P. Ultrafast graphene photodetector. Nature Nanotechnology. 2009;**4**(12):839

[141] Tao L, Chen Z, Li X, Yan K, Xu JB. Hybrid graphene tunneling photoconductor with interface engineering towards fast photoresponse and high responsivity. npj 2D Materials and Applications. 2017;**1**(1):19

[142] Xu X, Gabor NM, Alden JS, van der Zande AM, McEuen PL. Photothermoelectric effect at a graphene interface junction. Nano Letters. 2009;**10**(2):562-566

[143] Lopez-Sanchez O, Lembke D, Kayci M, Radenovic A, Kis A. Ultrasensitive photodetectors based on monolayer MoS_2. Nature Nanotechnology. 2013;**8**(7):497

[144] Konstantatos G. Current status and technological prospect of photodetectors based on two-dimensional materials. Nature Communications. 2018;**9**(1):5266

[145] Konstantatos G, Badioli M, Gaudreau L, Osmond J, Bernechea M, De Arquer FP, et al. Hybrid graphene–quantum dot phototransistors with ultrahigh gain. Nature Nanotechnology. 2012;7(6):363

[146] Lee Y, Kwon J, Hwang E, Ra CH, Yoo WJ, Ahn JH, et al. High-performance perovskite–graphene hybrid photodetector. Advanced Materials. 2015;**27**(1):41-46

[147] Chen Z, Cheng Z, Wang J, Wan X, Shu C, Tsang HK, et al. High responsivity, broadband, and fast graphene/silicon photodetector in photoconductor mode. Advanced Optical Materials. 2015;**3**(9):1207-1214

[148] Liu Y, Wang F, Wang X, Wang X, Flahaut E, Liu X, et al. Planar carbon nanotube–graphene hybrid films for high-performance broadband photodetectors. Nature Communications. 2015;**6**:8589

[149] Park J, Ahn YH, Ruiz-Vargas C. Imaging of photocurrent generation and collection in single-layer graphene. Nano Letters. 2009;**9**(5):1742-1746

[150] Miao X, Tongay S, Petterson MK, Berke K, Rinzler AG, Appleton BR, et al. High efficiency graphene solar cells by chemical doping. Nano Letters. 2012;**12**(6):2745-2750

[151] An X, Liu F, Jung YJ, Kar S. Tunable graphene–silicon heterojunctions for ultrasensitive photodetection. Nano Letters. 2013;**13**(3):909-916

[152] Chernikov A, Berkelbach TC, Hill HM, Rigosi A, Li Y, Aslan OB, et al. Exciton binding energy and nonhydrogenic Rydberg series in monolayer WS_2. Physical Review Letters. 2014; **113**(7): 076802

[153] Tsai DS, Liu KK, Lien DH, Tsai ML, Kang CF, Lin CA, et al. Few-layer MoS_2 with high broadband photogain and fast optical switching for use in harsh environments. ACS Nano. 2013;**7**(5):3905-3911

[154] Yu WJ, Liu Y, Zhou H, Yin A, Li Z, Huang Y, et al. Highly efficient gate-tunable photocurrent generation in vertical heterostructures of layered materials. Nature Nanotechnology. 2013;**8**(12):952

[155] Furchi MM, Polyushkin DK, Pospischil A, Mueller T. Mechanisms

of photoconductivity in atomically thin MoS_2 Nano Letters. 2014;**14**(11): 6165-6170

[156] Massicotte M, Schmidt P, Vialla F, Schädler KG, Reserbat-Plantey A, Watanabe K, et al. Picosecond photoresponse in van der Waals heterostructures. Nature Nanotechnology. 2016;**11**(1):42

[157] Ross JS, Klement P, Jones AM, Ghimire NJ, Yan J, Mandrus DG, et al. Electrically tunable excitonic light-emitting diodes based on monolayer WSe_2 p–n junctions. Nature Nanotechnology. 2014;**9**(4):268

8

Study of Morphological, Electrical and Optical behaviour of Amorphous Chalcogenide Semiconductor

Mohsin Ganaie and Mohammad Zulfequar

Abstract

Amorphous chalcogenide semiconductor plays a key role in search for novel functional materials with excellent optical and electrical properties. The science of chalcogenide semiconductor (CS) show broad spectrum of soluble alloy and a wider band gap device that access the optimal energy bandgap. The electronic properties of these alloys can be tuned by controlling the proportion of (S, Se, Te). The chalcogenide semiconducting (CS) alloys are promising candidates because of low band gap (1.0–1.6 eV) and high extinction coefficient in the visible region of solar spectrum. The band structure of amorphous semiconductor governed the transport properties and evaluates various factors such as Tauc gap, defect states, mobility edges. In the extended and localized state of amorphous semiconductor an electron goes various transition, absorption/emission, transport which is due to drift and diffusion under DC electric fields. CS, including sulfides, selenides, and tellurides, have been broadly utilized in variety of energy conversion and storage devices for example, solar cells, fuel cells, light-emitting diodes, IR detector, Li/Na-ion batteries, supercapacitors, thermoelectric devices, etc. Here, we report various morphological electrical, structural, and optical properties of InSeS thin films prepared by Melt Quenching thermal evaporation technique.

Keywords: amorphous chalcogenide semiconductor, electrical properties, optical properties

1. Introduction

Solid can be found to prepare either in crystalline state (periodic) or in non-crystalline (disordered) state on the basis of their atomic-scale structure. The crystal has periodicity in its atomic structure and exhibit a property called long-range order or translational periodicity; positions repeat in space in a regular array to an infinite extend, as shown in **Figure 1a**. While non-crystalline state is disordered structure, does not possess long range order (or translational periodicity) as indicated in **Figure 1b** [1, 2], and also do not exhibit any discrete diffraction pattern [3, 4].

The non-crystalline solid is further divided into glassy and amorphous, which does or does not have glass transition temperature respectively [5]. The terms amorphous and non-crystalline are synonymous under this definition. The term glassy have the same structural meaning; and has been reserved for amorphous solid, which have been prepared by melt-quenching and exhibits a glass transition. Whereas amorphous refer to non-crystalline material that can only be prepared in thin film form on glass substrate, which are sufficiently cool to prevent crystallization. Glassy materials are sub-set of amorphous solids; all glasses are amorphous but all amorphous are not necessarily glasses [6, 7].

Furthermore, the structure of amorphous materials is neither periodic as that in crystal nor completely random as that in gas. Short range order similar to crystal-line materials exist in these disordered solids. The amorphous solid extend a vast range between completely perfect and completely random structure. An amorphous material does not take a definite phase, but is meta-stable or quasi-equilibrium.

The thermodynamic of liquid–solid transition is that when liquid is cooled, two events may occur either crystallization may take place at melting point Tm or the liquid will become supercooled for the temperature below Tm. As the temperature of supercooled is decreases the viscosity increases and ultimately reaches a value, which is typically of solid ($\approx 10^{14.6}$ poise), at this stage liquid is said to be in glassy state. These changes can be observed by monitoring the change in its volume as a function of temperature as shown in **Figure 2**. The crystallization process is

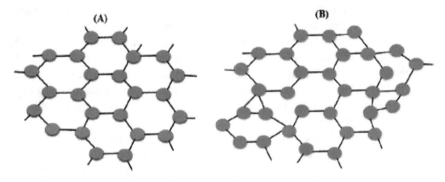

Figure 1.
Comparison of short-range order in crystal (A) and non-crystal (B).

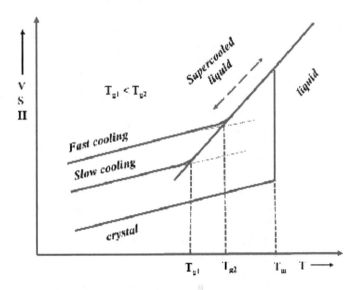

Figure 2.
Schematic illustration of change in volume with temperature, effect of cooling rate on glass transition temperatures is also shown.

manifested by an abrupt change in the thermodynamic variable at melting temperature Tm, which is discontinuous process. Whereas glass formation is characterized
by a gradual break in the slope is a continuous process, and the region where this
transition occur is known as glass transition temperature Tg. Other thermodynamic
variables such as entropy (S) and enthalpy (H) exhibits a similar behavior with
change in temperature. In the temperature interval between Tm and Tg2, the liquid
is referred to as the undercooled or supercooled [8]. The glass transition temperature Tg is different for different cooling rate, it is found that, slower the rate of
cooling, the larger is the region for which the liquid may be supercooled. Thus, the
glassy transition temperature of a particular material is not an intrinsic property,
but depends on the thermal history of the materials. According to the nucleation
theory, a liquid with a high viscosity between Tm and Tg typically exhibits a high
glass forming ability with a low critical cooling rate for glass formation [9]. Thus,
the glass formation is a matter bypass crystallization and nearly all materials can be
prepared in a glassy state, if cooled fast enough [10, 11].

The electronic density of states in crystal is well defined by block wave function
which extends throughout the solid; therefore the wave vector k is helpful in describing the electronic behavior and band gap in crystalline semiconductor. However, the
definition of the band gap is not as straightforward for amorphous semiconductors,
since there is no well-defined density of states, but there exist a continuous range of
states with in the energy level. Due to the disorderness which causes the localization
of electron, i.e. the wave function is confined within a small volume. Thus, the Shape
of wave function of electron in crystal and non-crystal are extended and localized
respectively. Consequently, their charge mobility in amorphous is smaller than that in
crystal. The lattice vibration are also extended and localized in crystal and non-crystal,
which give higher and lower thermal conductivities. This characteristic difference
between the electronic band structure of crystalline and non-crystalline solids plays an
important role in the electrical and optical properties. Also, their conductivity cannot
generally be increased appreciably by the incorporation of impurities, because the
impurity atom can satisfy only the valence requirement. Large fluctuations exist in the
local arrangement of atoms, which can accommodate an impurity without leading to
an extra electron or hole, thus exerting little effect on electrical properties. Due to
disorderness in amorphous semiconductors it was believed that there is no use of these
materials for technological applications [12, 13].

In 1955, Kolomiets and Gorunova from Ioffe Institute USSR [14], who discover
the first semiconducting glass $TlAsSe_2$ [15] in the system $Tl_2Se.Sb_2Se-TlSe.AsSe_3$.
The investigation of Kolomiets, large number of chalcogenide glasses with varying
Stoichiometry and composition can be synthesized. Later, in 1968, Ovshinsky and
his co-workers [16] from Energy Conversion Devices discovered the memory and
switching effects exhibited by some chalcogenide glasses. This shows that there is
transition from crystalline phase to amorphous phase and back by switching the
electric pulse, they called these materials as phase change materials. The switching
effect consists of a sharp transition from high resistivity to low resistivity of the
material, when an enough large electric field is applied. The observed switching
process was reversible and could be repeated many times. Kolomiets and Nazarova
[17] generalized that the amorphous semiconductors are always intrinsic and their
conductivity cannot be affected with doping.

Ovshinsky, Fritzsche et al., and Kastner reports indicate that drastic change can
be achieved in conductivity and activation energy by suitable preparation technique
[18, 19]. Amorphous semiconductor can serve as stable infrared transmitting
materials, which can use as optical materials for military purposes as lenses in night
goggles, photovoltaic solar cells with high efficiencies [20, 21]. They have a wide
range of photo-induced effects having a majority of applications. After this a new

effect has been observed such as photo doping radiometry, holography, optical memory effect having a huge potential application for imaging and electrophotography (Xerography).

More recently, the thin film form of these materials have gained much interest due to their structural, morphological and transport properties for their considerable use in various technological purposes. II–VI group semiconductor materials, plays a prominent part in the modern material science and technology and also for fabrication of large area arrays sensors, interface items, photo conductors, anti-reflector coating, IR detectors, optical fibers, etc. [22, 23]. The binary alloys of amorphous semiconductor are technologically important materials due to their direct and rather large gap [24]. These materials can be engineered for better application purpose by varying the composition parameter.

2. Classification and preparation of amorphous semiconductors

Amorphous semiconductor can be classified as tetrahedral-bonded Si like a material such as amorphous hydrogenated silicon (a-Si-H) and related alloys like Si-Ge:H which is four fold coordinated and leads to the symmetrical bonding and also result in the formation of rigid structure having a wider application in thin film transistor, photovoltaic, sensors etc. The second class includes Chalcogenide glasses which have at least one chalcogenide atom, VI group of periodic table (S, Se, and Te) as their major constituent. The four-fold coordination of Si leads to the forma-tion of highly symmetrical and rigid structures. On the other hand, twofold coor-dination in chalcogens is highly asymmetrical and the structure gives rise to greater degree of flexibility. Thus, the structure of chalcogenide is more complex than that of tetrahedral-bonded Si like material. Chalcogenide semiconductors consist of chain or ring like structures and have lone pair, so they are also called lone pair semiconductor. The oxygen is excluded in chalcogenide group which form a special class of material called as semiconducting oxide glasses. The present monograph is dedicated to the second group of materials, i.e., amorphous chalcogenide semicon-ductor. The classification of amorphous semiconductor is shown in **Table 1**.

The electronic density of states in crystal is well defined by block wave func-tion which extends throughout the solid; therefore the wave vector k is helpful in describing the electronic behavior and band gap in crystalline semiconductor. However, the definition of the band gap is not as straightforward for amorphous semiconductors, since there is no well-defined density of states, but there exist a continuous range of states with in the energy level. Due to the disorderness which causes the localization of electron, i.e., the wave function is confined within a small volume. Thus, the shape of wave function of electron in crystal and non-crystal are extended and localized respectively. Consequently, their charge mobility in amorphous is smaller than that in crystal. The lattice vibration are also extended and localized in crystal and non-crystal, which give higher and lower thermal conductivities. This characteristic difference between the electronic band structure of crystalline and non-crystalline solids plays an important role in the electrical and optical properties. Also, their conductivity cannot generally be increased appre-ciably by the incorporation of impurities, because the impurity atom can satisfy only the valence requirement. Large fluctuations exist in the local arrangement of atoms, which can accommodate an impurity without leading to an extra electron or hole, thus exerting little effect on electrical properties. Due to disorderness in amorphous semiconductors it was believed that there is no use of these materials for technological applications.

Tetrahedrally bonded semiconductor		Amorphous chalcogenide semiconductor		Semiconducting oxide glasses
C	InSb	S	Cd-Se-S	V_2O_5-P_2O_5
Si	GaAs	Se	In-Se-S	MnO-Al_2O_3-SiO_2
Ge	Si-Ge:H	Te	MWCNT/SeTe	V_2O_5-PbO-Fe_2O_3

Table 1.
Classification of amorphous semiconductor.

Amorphous solid can prepared from all phases of matter, solid, liquid or vapor as starting materials, but the deposition from vapor or liquid phases are more considerable. Since amorphous solid are less thermodynamically stable (it possesses a greater free energy). The preparation of amorphous material can be regarded as the addition of excess of free energy. Many preparation techniques are possible, depending on what kinds of materials are required for research and/ or applications. Melt quenching is one of the oldest techniques use to prepare amorphous solid especially, bulk glasses. When thin films are required different techniques are used such as, thermal evaporation, chemical vapor deposition, Gel desiccation, Electrolytic deposition, Sputtering, Chemical reaction, Reaction amorphization, Irradiation, Shock-wave transformation, Shear amorphization, Glow discharge decomposition. In this thesis, thermal evaporation technique is used for the preparation of amorphous semiconductors in thin film form under a vacuum of 10^{-6} Torr.

3. Band models of amorphous semiconductors

The significance of the band theory is used to explain or predict the proper-ties of materials like electrical and optical. The band itself denotes the number of electronic states per unit energy per electron. In case of crystalline solid, the band diagram is represented in **Figure 3a**, which shows two results, Sharpe definite edges where the density of states decreases to zero and the wave function is extended throughout the solid. Band theory of amorphous semiconductor was first explained by Mott [25]. Based on Anderson's theory [26] sufficiently large amount of disorder results in the localization of all electronic states in the band. These localized states do not occupy all the energy states in the band, but form a tail above and below the band. The localized and delocalized regions (tail states) can be found separately by the mobility edge.

Several models have been proposed to account for the high density of defect level and also for the band structure of amorphous semiconductors, where the energy distribution and density of states (DOS) is main concerned. The existence of local-ized states in the tails of valence band and conduction band and of a mobility edge separating the extended states from the localized states.

3.1 Cohen-Fritzsche-Ovshinsky (CFO) model

In 1969, the energy state described by Cohen, Fritzsche and Ovshinsky [27] assumed that tail states exist across the gap that separates the localized states from the extended ones. In their energy band model, the disorder is sufficiently high that causes conduction and valence band tails overlap in the mid gap leadings to appreciable density of states in the overlapping band, **Figure 3b** shows CFO model.

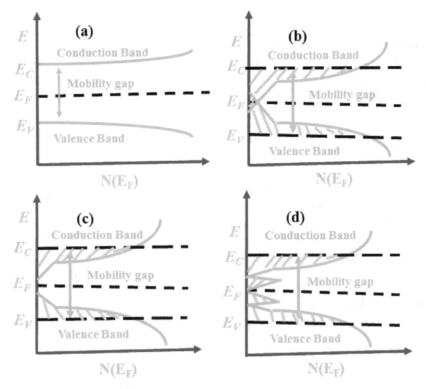

Figure 3.
Schematic density of states diagrams for amorphous semiconductors. (a) Crystalline semiconductor, (b) Cohen-Fritzsche-Ovshinsky model, (c) Davis-Mott model showing a band of compensated levels near the middle of the band gap, and (d) Marshall and Owen model [19].

There is a sharp mobility edges in each band that separate the extended states from the localized states. As a result of band overlapping redistribution of electron must take place which form filled states in the conduction band tail, which are negatively charged and empty states in the valence band which are positively charged. The CFO model was especially proposed to understand the electrical switching proper-ties in chalcogenide glasses. The primary objection to the CFO model is the high transparency observed in the chalcogenide glass, below a well-defined absorption edge.

3.2 Davis and Mott (DM) model

Davis and Mott [28] suggest a model for electrical conduction in amorphous solid in which tail of localized states should be narrow and should extend a few tenths of an electron volt into the forbidden gap. Defects in the random network such as dangling bonds, vacancies etc., give rise to the band compensation level near the middle of the gap. The central compensated band may be spilt in to a donor and accepter band results in the pinning of Fermi level at center. It was suggested by Mott that as the transition from extended to localized states, mobility drop by several order of magnitudes producing mobility edge. **Figure 3c** shows Davis Mott model, here E_C and E_V represent the energies, which separate the range where the state are extended and localized. The energies at E_C and E_V act as pseudo gap and define as a mobility gap.

Deviation from the 8-N rule will raise the topological defects, unsatisfied, or broken, or dangling bond, these kinds of defects should presents in large quantities of amorphous network. Anderson postulated [29] that negative effective correlation energy, or negative –U, centers in these amorphous chalcogenide semiconductors is due to the strong electron–phonon coupling.

Marshall-Owen [30] modified the Davis-Mott model suggested that the position of Fermi level is determined by well separated band of donor and acceptor level with some overlapping in the upper and lower halves of the mobility gap respectively **Figure 3d**. The self-concentration of the donor and acceptor takes places in such a way that Fermi-level remain in the mid of the gap.

3.3 Street and Mott model

Based on Anderson's idea, Street and Mott [31] proposed the model of charged defects states, instead of neutral dangling bonds in the band gap region around the Fermi level in amorphous chalcogenide materials. A bond is transferred from one chalcogenide atom to another resulting in two defects: one negatively charged, and other positively charged, labeled respectively (D^- and D^+), D stand for atom having dangling bond. These (D^- and D^+) exist in a more stable way than neutral D^0. It is proposed that the reaction is isothermal, and exothermic in nature. These charged defects states are electron-spin resonance (ESR) inactive, because all electrons are paired. This model assumes that electron–phonon interaction makes electron pair-ing energetically favorable at defects.

$$2\,D^0 \rightleftarrows D^+ + D^- \tag{1}$$

This model successfully explains variety of external induced phenomena in chalcogenide semiconductor such as photoconductivity, thermal process, luminance etc.

3.4 Kastner Adler and Fritzsche (KAF) model

Later on Kastner, Adler and Fritzsche [32] proposed a model know a valence-alternation pair (VAP) for the formation of over-coordinated defects through the involvement of lone-pair electron. Amorphous chalcogenide are therefore called lone-pair (LP) semiconductors. The tetrahedrally bonded amorphous chalcogenide semiconductor only two of the three p orbitals can be utilized for bonding, which split into the bonding (σ) and anti-bonding (σ^*) molecular states, that are subsequently broadened into the valence and conduction bands respectively. One normally finds chalcogens in twofold coordination. This leaves one non-bonding electron pair, termed lone pair (LP), these lone-pair, electrons form a band near the original p-state energy.

Emin [33], in 1975, proposed that charge carriers in amorphous semiconductor may enter self-trapped states (small-polarons) which is due to the polarization in the surrounding lattice. Emin suggest that the presence of disorder tends to slow down the carriers, which may leads to the localization of carrier in that states. Emin explain various experimental data of amorphous semiconductors such as thermos-physics, Hall-mobility, DC conductivity in the framework of exciting polarons theories.

4. Preparation of amorphous semiconductors

4.1 Melt-quenching technique

Melt quenching is one of the oldest technique use to prepare amorphous solid especially bulk glasses, In this method a melt is rapidly cooled by simply turning off the furnace, or rapid cooling in ice cooled water or liquid nitrogen depending upon the required cooling rate. Upon cooling a liquid below its melting point it will crystallize to form glasses. Amorphous material is formed by the process of continuous hardening (i.e., increase in viscosity) of the melt. An essential feature of glass formation from melt is fast cooling which prevent nucleation and growth [34].

The constituent elements having high purity (99.999%) with desired compositional ratio, according to their atomic mass ratio have been weight using electronic balance least count (10^{-4} gm.) were sealed in Quartz ampoules and maintaining a vacuum of order (10^{-6} Torr). The flow chart for the preparation of amorphous semiconductor is given in **Figure 4**.

The sealed ampoules were then placed in a rocking furnace where continuous heating is done at rate of 4–5 K/min to ensure the homogenization of the melt. In case of amorphous semiconductors materials, an ice cooled water is used to produce the glassy state. The quenched samples were removed by breaking the quartz ampoules. The obtained shiny bulk samples are grinded into fine powder using pastel and mortal.

4.2 Thermal evaporation technique

This technique is mostly use for the preparation of materials in thin film form, in which a material to be evaporated is placed in a boat. Schematic diagram of vacuum chamber is shown in **Figure 5**. Vacuum chamber containing molybdenum boat, where the source material is placed. The boat is connected with high voltage potential difference through which boat can be heated till the material is evaporated. The material which is placed in the boat must have low melting point than the molybdenum, which can evaporate easily and uniformly spread on substrate.

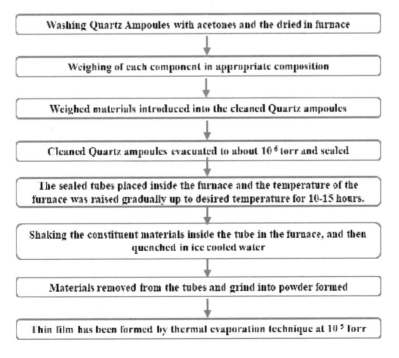

Figure 4.
Design flow chart for preparation of bulk amorphous samples.

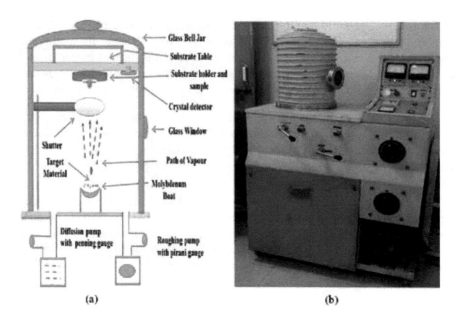

Figure 5.
(a) Schematic diagram of thermal evaporation technique (b) photograph of thermal evaporation coating unit to be used.

The evaporation rate can be controlled with the help of shutter, which is located above the boat and the thickness of the deposited film can be controlled with the help of single crystal thickness monitor, which is mounted near the sample holder.

The evaporation is performed in a vacuum to avoid the contamination; chamber is evacuated by diffusion pump and backed by rotary pump. The rotary pump used to evacuate the chamber to a pressure within the operating range of a diffusion pump; this is known as rough vacuum. Once the chamber reached at the vacuum equal to 10^{-3} Torr. An oil diffusion pump which can produce a vacuum of 10^{-6} Torr in the chamber must be maintained for the evaporation of materials. The pirani gauge and penning gauge is used to measure the pressure. The distance between the substrate and the target materials should not be more than 15 cm. Substrate temperature is important parameter, if higher the temperature of the substrate, the material will crystallize, room temperature is quite enough for the amorphization of the materials.

5. Basic properties of amorphous semiconductors

5.1 Structural properties

Amorphous semiconductors a new discovery in 1954 by Kolomiets and Goryunova shows that these materials possess semiconducting properties, first tried group VI elements (S, Se, Te) in place of oxygen elements, which was used to prepare oxide glasses. Following the pioneering work by Ioffe and Regel [35], Ioffe realized that amorphous semiconductor attracts a lot of attention due to their unique properties. The amorphous materials could behave as semiconductor and the band gap depends on the short range order rather than long range order. Also amorphous material possesses a large number of defects, which in turn creates a large number of localized states. In crystalline materials, the band structure is completely determined by translation symmetry in its structure, since all equivalent interatomic distance and bond angles are equal. But in amorphous material bond angle and bond lengths are nearly equal. The short order range on atomic scale is usually similar to that in corresponding crystal. Little variation in the bond length and bond angle and other imperfection results in broadening of the band edges.

There have been many discussions on the type of structure that has been developed for amorphous materials depending on their chemical nature. In order to obtain complete structure of amorphous semiconductors, following experimental tools are presented. The overall aspect of the chemical structure that constitute short range order and long range order are obtained by using X-ray, electron and neutron beams. Extended X-ray absorption fine structure (EXAFS) is a probe particularly sensitivity for investigating the local structure around the atoms. The morphology of the samples and the inhomogeneity present in the form of phase separation and voids can be examined by transmission electron microscopes (TEM) and scanning electron microscope (SEM). The Raman and infra-red spectroscopy are the mean suitable for examine the degree of disorder and chemical bonding. The medium range of amorphous network can be well understood by small angle scattering. Differential thermal analysis is used to study the variation in structures with temperature.

5.2 Electrical properties

The band states that exist near the middle of the gap arises the defects such as dangling bond, interstitial etc. These structural defects play an important role in electrical properties. Since all the electronic wave function in crystal are extended and in non-crystal the wave function of electron and hole are localized. In consequences the electron/hole mobility and the absorption in non-crystal become smaller that in corresponding crystal. Amorphous semiconductor basically has low thermal and electrical conductivity, since their band gap is large and the Fermi level lie within the gap. When talk about the electronic properties of amorphous semiconductors, the electrical conductivities at room temperature are in the range of 10^{-2} to $10^{-15}\ \Omega^{-1}\ cm^{-1}$. Conductivity is directly proportional to the electron/hole transport properties in the materials.

Electrical properties are primarily determined by electronic density of states, the electronic properties yielded valuable ideas like mobility edge, charge defects and activation energy. Amorphous semiconductors exhibit smaller conductivity than the corresponding crystal. Generally there are two type of conduction, band conduction (DC conduction) and hopping conduction. The band conduction occurs when the carriers would excite beyond the mobility edges into non-localized states.

The charge transport properties significantly depends on the energy spectrum Ev and Ec denote the mobility edges for the valence and conduction bands, respectively. Electronic states in the mobility gap between these energies are spatially localized. The states below Ev and above Ec can be occupied by delocalized holes and electrons respectively. The nature of charge carrier transport gets altered, when the carrier crosses the mobility edges, Ec and Ev. The transport above Ec is the band conduction type for electrons and the transport below Ev is band conduction for holes. Mostly the conductivity in amorphous semiconductor at higher temperatures is dominated by band conduction but, at lower temperatures, hopping conduction dominates over band conduction. **Figure 6** shows the different types of conduction in amorphous semiconductor.

Davis and Mott predict the DC conductivity of amorphous semiconductor, depending on the temperature range, three conduction mechanism are possible [36].

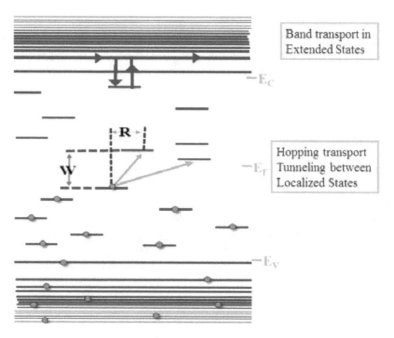

Figure 6.
Different types of conduction in amorphous semiconductor.

5.3 Optical properties

Amorphous semiconductor are the promising candidate for the photonic application owing to their interesting optical properties like high absorption coefficient, high efficiency of radiative recombination, high photosensitivity and nearly matching band gaps with visible region of solar spectrum. The design and realization of optical components based on these materials requires detailed information on their optical properties like electronic band structure, optical transition and relaxation mechanism and also a huge amount of information about their structure, opto-electronic behavior, transport of charge carriers, etc. The band gap represents the minimum energy difference between the top of valence band and the bottom of conduction band. However the minimum of the conduction band and maximum of the valence band occur at the same value of momentum, the energy difference is released as photon, this is called direct band gap semiconductor. In an indirect band gap semiconductor, the maximum energy of the valence band occurs at different value of momentum to the minimum of the conduction band energy (**Figure 7**). Both the transition direct and indirect gives rise to the frequency dependence of absorption coefficient near the absorption edge.

In amorphous semiconductors there are three distinct regions for optical absorption curve.

a. High absorption region ($\alpha \geq^{4} 10$ cm $)^{-1}$ which gives the optical band gap between the valence band and conduction band, the absorption coefficient has the following frequency dependence [37].

$$\alpha h\upsilon = A \left(h\upsilon - E_g \right)^n \tag{2}$$

where A, E_g, υ, h are constant, optical band gap, frequency of incident radiation and plank's constant respectively. The index 'n' is associated with the type of

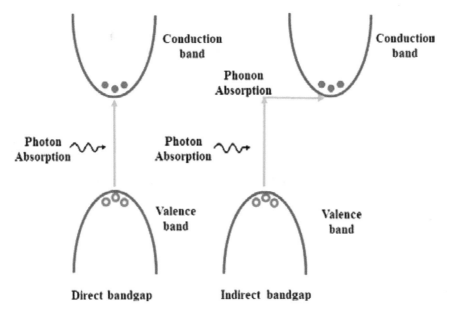

Figure 7.
Schematic representation of direct and indirect band gap.

transition, which takes the values n = 1/2, 2, 3 and 3/2 for direct allowed, indirect allowed, indirect forbidden and direct forbidden transitions, respectively.

b. Spectral region with ($\alpha = 10^2$–10^4 cm^{-1}) is known as Urbach's exponential tail region. In this region the optical transition takes places between localized tail states and extended tail states. The absorption coefficient can be fitted by an Urbach law.

$$(Ah\nu)^n = \alpha_0 \exp\left[^{h\vartheta/E_u}\right] \tag{3}$$

The exponential factor *Eu* indicates width of band tails of the localized states and is independent of temperature. At low temperature has a value of about 0.05–0.08 eV for most of amorphous semiconductor [38].

c. The region ($\alpha \leq 10^2$ cm^{-1}) involves the low energy absorption region, defects and impurities originates in this region [39].

Extensive researches on amorphous Chalcogenide semiconductors attract the attentions of researchers and engineers begins nearly at the end of 1990s, when significant change in the physical, electrical, phase, physicochemical and structural properties induced by light of appropriate energy and intensity has been investigated. The properties include photoconductivity, photovoltaic, photoluminescence, and non-linear optical phenomena that are connected with purely electronic effects.

a. At high temperature, conduction may take place by carriers into the excited states beyond the mobility edges, electron in the conduction extended states and hole in the valence extended states. This yield an activated type temperature dependence for the conductivity of electron as [40].

$$(\sigma_{dc}) = \sigma_0 \exp.\left[-\frac{Ec - Ev}{KT}\right] \tag{4}$$

where Ec – Ev = ΔE are the activation energy, the electron at or or above Ec move freely, while below it move through an activated hopping. σ_0 is the pre-exponential factor.

b. At intermediate temperatures, conduction occurs by thermally assisted tunneling of carriers excited into localized gap state near the mobility edges; at E_A and E_B.

The thermal assisted tunneling process means localized electron jump from site to site with the exchange of phonon, this process is called hopping conduction. If the conductivity is carried by electron then conduction will be [41]

$$(\sigma_{dc}) = \sigma_1 \exp.\left[-\frac{E - Ev - \Delta W1}{KT}\right] \qquad (5)$$

where $\Delta W1$ is the hopping energy in addition to the activation energy (Ec – Ev) needed to raise the electron to the appropriate localized states at energy E. Where $E = E_A$ and E_B is the extremity of the conduction and valence band tail. σ_1 is expected to be less than σ_0 by a factor of 10^2 to 10^4.

c. At low temperature, conduction occurs by thermally assisted tunneling of carriers between the localized states near the Fermi energy E_F.

The carriers move between the states located at E_F via phonon assisted tunneling process, which is analogous to impurity conduction observed in heavily doped and highly compensated semiconductors at low temperatures, the conductivity is given by [42]

$$(\sigma_{dc}) = \sigma_2 \exp\left[-\frac{\Delta W2}{KT}\right] \qquad (6)$$

where $\sigma_0 < \sigma_1$, ΔW_2 has the same physical meaning as $\Delta W._1$ Since the density of states near E_F and the range of their wave-functions is probably smaller near E_F than near E_A or E_B.

6. Results

Bulk ternary $In_4Se_{96-x}S_x$ (where x = 0, 4, 8, 12) semiconductor were prepared by melt-quenching method from high purity (99.999%) reagents. **Figure 8** show the amorphous nature of the prepared samples was carried out by a Rigaku Ultima IV X-ray diffractometer (XRD).

Morphological analysis were done by Scanning Electron Microscope (SEM), Scanning electron microscope is an influential procedure to learning the morphology, growth mechanism, particle size, shape and distribution of particles in the films. Surface morphology of thin films samples have been studied by scanning electron microscopy (FESEM) apparatus shown in **Figure 9**. We acquire SEM micrograph from Model: Σ igma by Carl Zeiss employed with Gemini Column (Patented technology of Corl Zeiss) which enables inlense secondary electron detection. SEM micrographs acknowledge the inclusive consistency of the thin film on glass substrate, which assert that the average grain size of the films increases from 130 to 160 nm and 190 nm with 4 and 8% of Sulfur doping, at 12% of sulfur doping it decreases to 135 nm. Thus, the improvement of development of particle size and the vanishing of small gaps between the grains after doping shows that the disorderness decreases with Sulfur concentration.

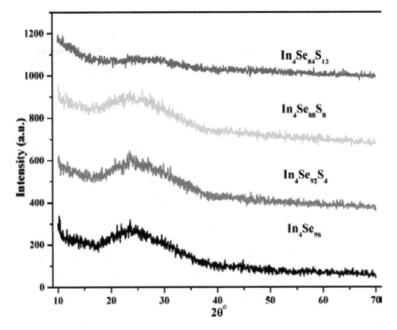

Figure 8.
The X-ray diffraction pattern of $In_4Se_{96-x}S_x$ (where x = 0, 4, 8, 12) thin films.

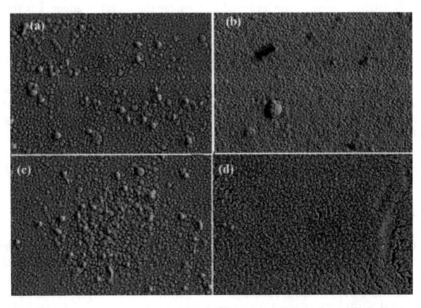

Figure 9.
SEM images of (a) In_4Se_{96}, (b) $In_4Se_{92}S_4$, (c) $In_4Se_{88}S_8$, and (d) $In_4Se_{84}S_{12}$ thin films. scale bars 200nm.

To measure the UV–Visible absorption and transmission spectra along with the energy band tail width and optical band gap in thin film of $In_4Se_{96-x}S_x$ (where x = 0, 4, 8, 12) has been recorded in the wavelength range of 190–1100 nm by UV–Vis spectrophotometer (Model: Comspec M550) is shown in **Figure 8**.

The relation between absorption coefficient and optical band gap is given by Tauc relation [32].

$$\alpha h\nu = A \left(h\nu - E_g \right)^n \qquad (7)$$

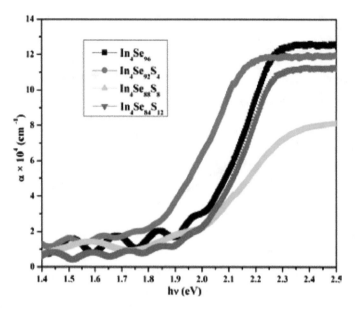

Figure 10.
Absorption coefficient (α) versus hυ of In₄Se₉₆₋ₓSₓ (where x = 0, 4, 8, 12) thin films.

where A is constant called the band tailing parameter, E_g is the optical band gap, which is situated between the localized states near the mobility edges, ν is the frequency of incident radiation and h is the plank's constant. The index 'n' is associated with the type of transition. The values of absorption coefficient (α) are in the range of 10^4 cm^{-1}. With growing the sulfur concentration, the band gap is lifted to higher energy. The upsurge in optical band gap on addition of sulfur in In-Se thin films may be due to reduction in disorderness of the system and also due to decrease in density of localized states. It is well known that during the deposition of thin film, amorphous semiconductor covers large number of defects states such as dangling bond composed with some saturated bond. The addition of sulfur in In-Se system consequences the reduction of unsaturated defects making a number of saturated bonds, which is accountable for the increase in optical band gap [43] (**Figure 10**).

7. Conclusions

Form the above studies, one can conclude Powder sample of InSeS chalcogenide semiconductors have been successfully prepared by Melt-quenching technique and then deposited by thermal evaporation technique on ultra-clean glass substrate show smooth coverage. XRD confirm the amorphous nature of the prepared thin film. The films are highly absorptive, furthermore, the optical band gap cover the complete visible region of solar spectrum. The structural, optical and morphologi-cal studies of InSeS thin film show that it is a promising material for PV application and have the potential to be transferred in industry applications.

Acknowledgements

Authors are also thankful to Department of Physics Jamia Millia Islamia, Central Research Facility (CRF) and Nanoscale Research Facility (NRF), IIT Delhi for necessary laboratory facilities. MG is thankful to DST-SERB for the award of NPDF (PDF/2017/000429).

Author details

Mohsin Ganaie[1*] and Mohammad Zulfequar[2]

1 Centre for Energy Studies, Indian Institute of Technology Delhi, Hauz Khas, New Delhi, India

2 Department of Physics, Jamia Millia Islamia, New Delhi, India

*Address all correspondence to: mohsin.ganaie@gmail.com

References

[1] Elliot SR. Physics of Amorphous Materials. 2nd ed. London: Longman; 1990

[2] Ioffe AF, Regel AR. Non-crystalline, amorphous, and liquid electronic semiconductors. Progress in Semiconductors. 1960;**4**:237-291

[3] Ganaie M, Zulfequar M. Structural and optical investigation of $Cd_4Se_{96-x}S_x$ (x= 4, 8, 12) chalcogenide thin films. Journal of Materials Science: Materials in Electronics. 2015;**26**:4816-4822

[4] Ganaie M, Zulfequar M. Structural, electrical and dielectric properties of CNT doped SeTe glassy alloys. Materials Chemistry and Physics. 2016;**177**:455-462

[5] Kugler S, Shimakawa K. Amorphous Semiconductors. Cambridge University Press; 2015. DOI: 10.1017/CBO978113 9094337

[6] Liu D, Xu L, Xie J, Yang J. A perspective of chalcogenide semiconductor-noble metal nano-composites through structural transformations. Nano Materials Science. 2019;1:184-197

[7] Chowdhury SR et al. Thermoelectric energy conversion and topological materials based on heavy metal chalcogenides. Journal of Solid State Chemistry. 2019;**275**: 103-123

[8] Schwarz RB, Johnson WL. Formation of an amorphous alloy by solid-state reaction of the pure polycrystalline metals. Physical Review Letters. 1983;**51**:415

[9] Jones GO. Glass. 2nd ed. London: Chapman and Hall; 1971

[10] Green MA et al. Solar cell efficiency tables (Version 49). Progress in Photovoltaics: Research and Applications. 2017;**25**:3-13

[11] Zhu M et al. Solution-processed air-stable mesoscopic selenium solar cells. ACS Energy Letters. 2016;**1**:469-473

[12] Yao Q et al. Nanoclusters via surface motif exchange reaction. Nature Communications. 2017;**8**:1555

[13] Pierre A et al. Charge-integrating organic heterojunction phototransistors for wide-dynamic-range image sensors. Nature Photonics. 2017;**11**:193-199

[14] Kolomiets BT. Vitreous semiconductors (II). Physica Status Solidi B. 1964;**7**:713

[15] Goryunova NA, Kolomiets BT. Electrical properties and structure in system of Selenide of Tl, Sb, and As. Zhurnal Tekhnicheskoi Fiziki. 1955; **25**:2669

[16] Ovshinsky SR. Reversible electrical switching phenomena in disordered structures. Physical Review Letters. 1968;**21**:1450

[17] Goryunova NA, Kolomiets BT. New vitreous semiconductors. Izvestiya Akademii Nauk, Seriya Fizicheskaya. 1956;**20**:1496

[18] Ovshinsky SR. In: Spear WE, editor. Proceedings of the International Conference on Amorphous Liquid Semiconductors. Edinburgh: CICL; 1977. p. 519

[19] Kastner PM. Prediction of the influence of additives on the density of valence-alternation centres in lone-pair semiconductors. Philosophical Magazine B. 1978;**37**:127-133

[20] Zallen R. The Physics of Amorphous Solids. New York, USA: Wiley; 1983

[21] Ovshinsky SR. Fundamentals of amorphous materials. In: Adler D, Schwartz BS, Steele MC, editors. Physical Properties of Amorphous Materials. Institute for Amorphous Studies Series. Vol. 1. New York: Plenum Press; 1985. p. 105

[22] Khan SA, Zulfequar M, Husain M. Effects of annealing on crystallization process in amorphous $Ge_5Se_{95-x}Te_x$ thin films. Physics B. 2002;**324**: 336-343

[23] Soumya Deo R et al. Structural, morphological and optical studies on chemically deposited nanocrystalline CdZnSe thin films. Journal of Saudi Chemical Society. 2014;**18**:327-339

[24] Cui S et al. Pressure-induced phase transition and metallization of solid ZnSe. Journal of Alloys and Compounds. 2009;**472**:294-298

[25] Mott NF, Davis EA. Electronic Process in Non-crystalline Materials. Oxford: Clarendon; 1979

[26] Anderson PW. Absence of diffusion in certain random lattices. Physics Review. 1958;**109**:1492

[27] Cohen MH, Fritzsche H, Ovshinsky SR. Simple band model for amorphous semiconducting alloys. Physical Review Letters. 1969;**22**:1065

[28] Davis EA, Mott NF. Conduction in non-crystalline systems V. Conductivity, optical absorption and photoconductivity in amorphous semiconductors. Philosophical Magazine. 1970;**22**:903

[29] Anderson PW. Model for the electronic structure of amorphous semiconductors. Physical Review Letters. 1975;**34**:953

[30] Marshall JM, Owen AE. Drift mobility studies in vitreous arsenic triselenide. Philosophical Magazine. 1971;**24**:1281-1305

[31] Street RA, Mott NF. States in the gap in glassy semiconductors. Physical Review Letters. 1975;**35**:1293

[32] Kastner M, Adler D, Fritzsche H. Valence-alternation model for localized gap states in lone-pair semiconductors. Physical Review Letters. 1976;**37**:1504

[33] Emin D. Aspects of the theory of small polarons in disordered materials. In: Le Comber PG, Mort J, editors. Electronic and Structural Properties of Amorphous Semiconductors. London/ New York: Academic Press; 1973. p. 261

[34] Elliott SR. Physics of Amorphous Materials. 2nd ed. Harlow/New York: Longman Science & Technical/Wiley; 1990

[35] Ganaie M, Zulfequar M, Journal of Physics and Chemistry of Solids. 2015;**85**:51-55

[36] Mott NF, Davis EA. Electronic Processes in Non-Crystalline Materials. New York: Oxford University Press; 1979

[37] Goswami A. Thin Film Fundamentals. New Delhi: New Age International Publishers; 1996

[38] Tauc J. Optical Properties of Amorphous and Liquid Semiconductors. New York: Plenum Press; 1974

[39] Wood DL, Tauc J. Weak absorption tails in amorphous semiconductors. Physical Review B. 1972;**5**:3144

[40] Nagels P. In: Brodsky MH, editor. Amorphous Semiconductors. Berlin: Springer Verlag; 1985

[41] Tanaka K, Maruyama E, Shimada T, Okamoto H. Amorphous Silicon. John Wiley & Sons; 1999

[42] Spear WE, Allan D, Le Comber P, Ghaith A. A new approach to the interpretation of transport results in a-Si. Philosophical Magazine B. 1980;**41**(4):419-438

[43] Ganaie M, Zulfequar M. Optical and electrical properties of $In_4Se_{96-x}S_x$ chalcogenide thin films. Journal of Alloys and Compounds. 2016;**687**: 643-651

9

Indenter Shape Dependent Dislocation Actives and Stress Distributions of Single Crystal Nickel during Nanoindentation: A Molecular Dynamics Simulation

Wen-Ping Wu, Yun-Li Li and Zhennan Zhang

Abstract

The influences of indenter shape on dislocation actives and stress distributions during nanoindentation were studied by using molecular dynamics (MD) simulation. The load-displacement curves, indentation-induced stress fields, and dislocation activities were analyzed by using rectangular, spherical, and Berkovich indenters on single crystal nickel. For the rectangular and spherical indenters, the load-displacement curves have a linear dependence, but the elastic stage produced by the spherical indenter does not last longer than that produced by the rectangular indenter. For a Berkovich indenter, there is almost no linear elastic regime, and an amorphous region appears directly below the indenter tip, which is related to the extremely singular stress field around the indenter tip. In three indenters cases, the prismatic dislocation loops are observed on the {111} planes, and there is a sudden increase in stress near the indenter for the Berkovich indenter. The stress distributions are smooth with no sudden irregularities at low-indentation depths; and the stress increases and a sudden irregularity appears with the increasing indentation depths for the rectangular and spherical indent-ers. Moreover, the rectangular indenter has the most complex dislocation activi-ties and the spherical indenter is next, while very few dislocations occur in the Berkovich indenter case.

Keywords: molecular dynamics simulation, nanoindentation, indenter shape, dislocation, stress distribution

1. Introduction

Nanoindentation is one of the most popular experiments to investigate the behavior of materials at the nanometer scale [1, 2]. Nanoindentation-probed materials' properties are usually different from their macroscopic counterparts and present different deformation mechanisms [3]. The atomistic models can give important qualitative information for the understanding of experimentally observed complex phenomena; especially, they have the indisputable advantage of analyzing the main physical process and micro-mechanisms (dislocation nucleation and evolution) [4–9]. Many studies have adopted atomistic simulations to probe and understand detailed deformation process and the possible impact factors involved in nanoindentation. For example, Chan et al. [10] studied the size effects of indenter nanohardness using atomistic simulations, and found that nanohardness was inversely proportional to the indenter radius. Imran et al. [11] studied the influence from the indenter velocity and size of the indenter using molecular dynamics (MD) simulations in Ni single crystals. Noreyan et al. [12] carried out MD simulations of nanoindentation of β-SiC to investigate the dependence of the critical depth and pressure for the elastic-to-plastic transition on indentation velocity, tip size, and workpiece temperature. Yaghoobi et al. [13] and Kim et al. [14] conducted MD simulations of nanoindentation on nickel single crystal and nickel bicrystal and discussed the effects of boundary conditions and grain boundary, respectively. Fu et al. [15] and Yuan et al. [16] investigated the effects of twin boundary on hardness, elastic modulus, and dislocation movements during nanoindentation by MD simulations. Fang et al. [17] carried out MD to find that both Young's modulus and hardness become smaller as temperature increases, and elastic recovery is smaller at higher temperatures during nanoindentation. Wu et al. [18] performed MD simulation to investigate the effects of the tempera-ture, loading and unloading velocities, holding time, and composition of Ni-Al alloys on the nanoimprinting lithography process. Hansson [19], Tsuru et al. [20], and Remington's et al. [21] studies indicated that the crystallographic orientation strongly influenced the hardness, load for pop-in formation, and mechanisms of plastic deformation using nanoindentation simulated by MD. Zhao et al. [22], Chamani et al. [23], and Chocyk et al. [24] also used MD simulation to study nanoindentation behaviors and microstructure features in metallic multilayers, and the effect of layer thickness on nanoindentation hardness was discussed. Furthermore, few studies also found that the geometry shape of indenter have a strong influence on the dislocation nucleation [25, 26], hardness [26, 27], and Young's modulus [27]. Generally, in the MD simulation of nanoindentation, the material properties response is influenced by some factors including specimen thickness, crystal orientation, grain size and boundaries, atomic potentials, boundary conditions, temperature, shape and size of indenter, and loading/ unloading rate of indentation, etc.

Although the atomic simulation of nanoindentation process and the influence of different factors have been considered comprehensively, most of them mainly con-sider the dislocation nucleation and deformation mechanisms of specimen, whereas seldom consider the evolutions of and microstructure and stress field of specimen caused by indenter tip geometry. Especially, the internal relationship between the microstructural evolution and the distribution of stress field during nanoindenta-tion has not been analyzed. Since the microstructure features and its evolutions in a localized region are closely related to the stress field in this region, the mechanical properties are strongly dependent on the microstructural features and their evolu-tions; and it is important to investigate the microstructure evolution and relation with the local stress field, which can be a great help for further understanding of the physical mechanisms during nanoindentation.

The objective of this contribution is to examine how the development of dis-location microstructures, stress distributions, and load-displacement curves are affected by complex stress sites from different shapes of indenters, to determine the relationships between dislocation microstructures and stress distributions during nanoindentation. Moreover, the effect of shapes of indenters on dislocation nucleation, and dislocation movements, stress distribution characteristics during nanoindentation of single crystal nickel are studied in detail.

2. Simulation procedure and stress calculation

In this work, an open source code LAMMPS [28] is used to carry out MD simulations and investigate the influences of indenter geometry on dislocation movements and stress distributions for a Ni single crystal nanoindentation. Three indenters with different geometries are employed: spherical, rectangular, and Berkovich indenters. The geometries of the models with three different shapes of indenters are shown in **Figure 1.** Both models are in the cubic orientation (i.e., X-[100], Y-[010], and Z-[001]), and the size of box ($X \times Y \times Z$) is 200 Å × 200 Å × 50 Å). A free boundary condition was applied on the top surface of z direction, while the last two layers (7.04 Å) at the bottom of z direction were frozen down, resembling a hard substrate. Periodic boundary conditions (PBC) were applied in the x and y directions, and the sizes of x and y directions, which are perpendicular to the z direction (indentation direction) were chosen to be large enough to avoid spurious effects of the PBC. In the present MD simulations, the embedded-atom-method (EAM) potential provided by Mishin et al. [29] was used, which has previously been successfully applied to simulate FCC single crystal nickel [30, 31]. Atoms in the indenter were kept fixed (indenter was assumed to be an infinitely rigid body), at the start of the simulation, all the models with different indenters are relaxed with the conjugate gradient method to reach a minimum energy state, and then the indenters are inserted into the free top surface of z direction at an average indentation speed of v = 0.25 m/s to a maximum penetration of 15 Å by using a displacement control, and this maximum depth is 30% of the specimen in z direction. When the indenter reaches the maximum depth, the indenter returns to the original position by unloading, hence the indentation process is completed. In addition, to reduce temporal local stress fluctuations, the systems are allowed to relax 4 ps after every increment of the indenters. Meanwhile, considering the time scale limitation of the MD simulation on a nanoindentation procedure, a very low temperature of 1 K is chosen to carry out all nanoindentation simulations, where the indentation rate does not play a significant role as long as it is less than the speed of sound [32]. In the present study, the atomic configurations and their evolutions are analyzed by common neighbor analysis (CNA) proposed by Honeycutt and Andersen [33], because CNA can provide an effective filtering method to classify

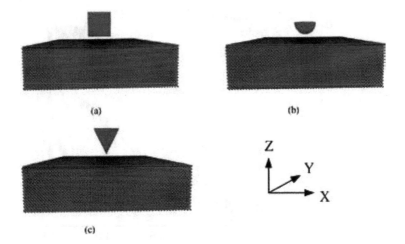

(a)   (b)

(c)

Figure 1.
The geometries of the models with three different shapes of indenters: (a) rectangular indenter, (b) spherical indenter, and (c) Berkovich indenter.

atoms and characterize the local structural environment in crystalline systems. Finally, the atomic configurations and their evolutions are observed by the visualization tools, AtomEye [34], which provides details of the microstructural evolution of nanoindentation.

To investigate the variation and distribution characteristics of atomic stress fields during nanoindentation, the atomistic stress definition is employed in this work. The stress tensor for an atom is calculated as a time averaged value over a number of time steps after relaxation of an indentation step, for atom α of the atomic stress tensor is defined as [35, 36]:

$$\sigma_{ij}^{\alpha} = -\frac{1}{V^{\alpha}}\left(M^{\alpha}v_i^{\alpha}v_j^{\alpha} + \sum_{\beta}F_i^{\alpha\beta}r_j^{\alpha\beta}\right) \qquad (1)$$

where V^{α} is the volume of atom α, M^{α} is atomic mass, v_i and v_j are the atom velocity components, and $F_i^{\alpha\beta}$ and $r_j^{\alpha\beta}$ is the force and distance between atoms α and β, respectively.

3. Simulation results and discussion

3.1 Effect of indenter shapes on load: Displacement curves

In order to compare the load-displacement curves of three different indenters, we gave three indenters penetrating into the same maximum indentation depth of 1.5 nm; the load-displacement curves for three different indenters are shown in **Figure 2**. As the indenters are −1 nm away from the contact surface of the specimen surface at the initial stage, the load-displacement is close to a straight line, and the load P is zero for three different shapes of indenter before the separation is approximately −0.5 nm. When the distance between the indenter and the contact surface of the specimen is about −0.5~0 nm, the load values are negative because of the repulsive force of the interaction between atoms, and the repulsive force is larger at the larger contact surface. The rectangular indenter has a larger contact surface than the spherical and Berkovich indenters, the repulsive force is biggest

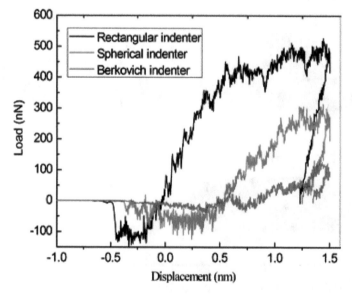

Figure 2.
Load-displacement curves for different shape indenters.

for the rectangular indenter and smallest for the Berkovich indenter (see **Figure 2**). After the indenter contact the surface of the specimen, the load gradually increases with the increase of the penetration depth, the load-displacement curve has a linear dependence, which means that at this stage, the indentation is performed in the linear elastic regime.

For a rectangular indenter, when the load-displacement curve reaches the maxi-mum load, the load keeps fluctuating around a constant value during the further displacement of the indenter, because of the constant cross section of the rectan-gular indenter; its further penetration into the crystal does not affect significantly the force acting on the indenter. The fluctuations are caused by the relaxation of stresses in the system, such as dislocations migration from the deformed sites. For the spherical indenter, similar to the rectangular indenter case, a quasi-elastic behavior is observed between inflection points; the force acting on the indenter gradually increases with the increase of the penetration depths, but does not last as long in increasing load periods. For a Berkovich indenter, the force acting on the indenter slowly increases with the increase of the penetration depths, the load drop events tend to be small or not noticeable.

Comparison with these three different indenters, the load-displacement curves show different peak loads, elastic-plastic behavior, and load drop events, which are related to a series of transitions of the dislocation structures under the indenta-tion site. To explain this phenomenon, we analyze the local stress distribution and microstructure evolution features of specimen under the indentation site during nanoindentation.

3.2 Effect of indenter shapes on stress distributions

Figure 3 shows atomic stress as a function of atom position along the X direc-tion for different shape indenters during nanoindentation. Here, four different nanoindentation sites are chosen to analyze the atomic stress distributions under the action of different indenters. For rectangular indenter, when indenter does not contact specimen surface [initial site (1) in **Figure 3a**], the atomic stress values are approximately zero along the X direction for three different shape indenters. When the indenter just contacts specimen surface [just contact site (2) in **Figure 3a**], near the contact point, the stress value is positive and relatively high. Whereas the speci-men surface at a distance from the contact point, the stress value is negative because the atomic interaction force is repulsive. As the indenter gradually penetrates into the specimen, the stress increases and a sudden irregularity appears. When indenter reaches the maximum indentation depth site (3), the stress value at this site is the highest, and a sudden increase occurs at the indenter site. When the indenter is fully unloaded to the initial site (4) in **Figure 3a**, comparison with the maximum inden-tation depth, the stress value decreases, but it is still high, which indicates that there are still dislocations in the unloaded specimens and cannot be restored completely. For spherical and Berkovich indenters, a similar trend is found with initially smooth stress distributions, becoming irregular at larger indentation depths. By comparing the stress distributions for three different shape indenters, we found that different indenters have similar stress distribution characteristics at different indentation depth sites, but the stress values are different. The rectangular indenter have the highest stress value (about 6.8 GPa), the spherical indenter is next (about 5.0 GPa), which also has a high stress value (nearly 5.0 GPa) for the Berkovich indenter at the same maximum indentation depth of 1.5 nm. Although the load is the smallest for the Berkovich indenter from the load-displacement curves in **Figure 2**, the contact area of the indenter is also the smallest, resulting in the stress value that is still high near the indenter. It is noteworthy that for the Berkovich indenter, except for the

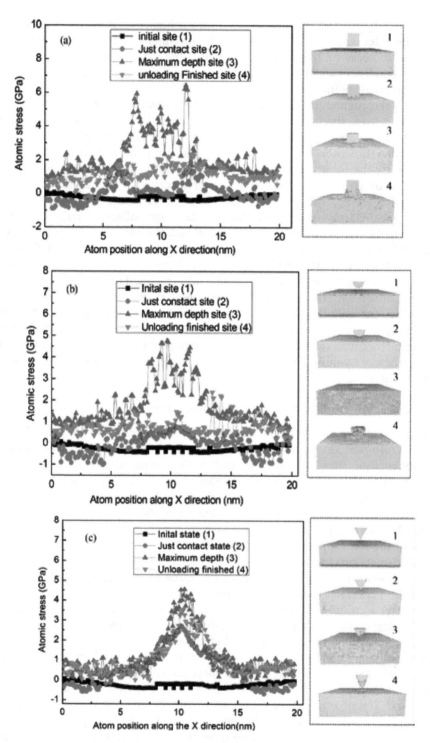

Figure 3.
Atomic stress as function of atom position along the X direction at four different nanoindentation sites for different shape indenters: (a) rectangular indenter, (b) spherical indenter, and (c) Berkovich indenter. The 1, 2, 3, and 4 inserted in figure indicate the indenter at four different nanoindentation sites: (1) initial site (indenter does not contact Ni single crystal specimen), (2) just contact site, (3) maximum depth site, and (4) unloading finished site.

high stress (nearly 5.0 GPa) near the indenter, the other parts of the specimen are very low, which is also the reason for the stress concentration of the sharp indenter. From **Figure 3**, we can see that a sudden increase in stress near the indenter for the

Berkovich indenter, while for the rectangular and spherical indenters, the stress distributions at low indentation depths are smooth with no sudden irregulari-ties, indicating a fully elastic response. At larger indentation depths, the stress distribution gets more and more irregular with locally very high stresses. The stress irregular distribution and the formation of pop-in are closely related to dislocation activities during nanoindentation.

3.3 Effect of indenter shapes on surface and internal microstructures of specimen

To understand the microstructure evolution of the specimen during nanoinden-tation process, the surface and internal microstructure characteristics of specimen for different shape indenters are presented at the maximum indentation depth site and unloading finished site, as shown in **Figures 4** and **5**, respectively.

Due to the maximum stress and irregular distribution occurring at the maxi-mum indentation depth site, and the dislocation activity is also the most complex at this site, From **Figure 4**, at the same maximum indentation depth of 1.5 nm, we can clearly see that the dislocation structure is the most complex for the rectangular indenter case, and the next for the spherical indenter case, whereas there are only few dislocations for the Berkovich indenter case. Compared to the dislocation activities of rectangular and spherical indenter cases, an amorphous region directly below the indenter tip is observed in the Berkovich indenter case, which we think that the extremely singular stress field around the indenter tip contributes to this uncommon observation.

When the unloading is finished, the indenter returns to its initial site, the stress cannot be completely restored and there are still a lot of dislocations on the surface of the specimen for the rectangular and spherical indenters cases, and these disloca-tions move slowly toward the surface and boundary of the specimen during unload-ing, forming the surface microstructure of the specimen as shown in **Figure 5a,b**, while in the Berkovich indenter case, there are almost no dislocations on the surface of specimen at unloading finished site (see **Figure 5c**).

Furthermore, for all three indenters cases at the same maximum indentation depth, the prismatic dislocation loops are mainly observed to nucleate on the {111} planes, as seen the red cycles in **Figures 4** and **5**, this observation in accor-dance with the predictions of dislocation theory in a face-centered cubic metal. In the rectangular indenter case, dislocation activities are found to be complex; a

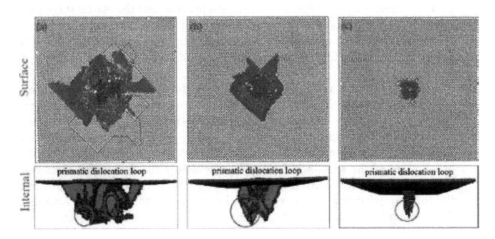

Figure 4.
The surface and internal microstructure features of specimens at maximum depth site for different shape indenters: (a) rectangular indenter, (b) spherical indenter, and (c) Berkovich indenter.

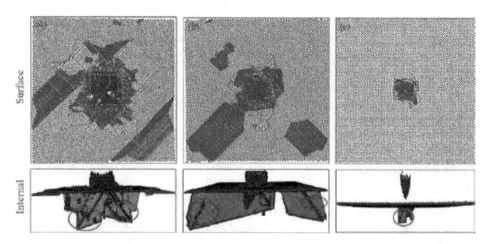

Figure 5.
The surface and internal microstructure features of specimen at unloading finished site for different shape indenters: (a) rectangular indenter, (b) spherical indenter, and (c) Berkovich indenter.

prismatic loop originating inside the specimen is observed and glided upward to the indenting surface. In the spherical indenter case, dislocation activities are still relatively complex, a prismatic dislocation loop originating from the surface of specimen is observed and glided downward to the boundary of specimen. In the Berkovich indenter case, dislocation activities are relatively scarce and the defect structure does not change significantly.

4. Conclusions

MD simulations are performed to investigate the influences of indenter geometry on the dislocation activities and stress distributions during nanoindentation on nickel single crystal. The stress field and dislocation activities induced by the change of indenter shape are analyzed. The major findings are as follows:

1. Load-displacement curves show different peak loads, elastic-plastic behavior, and load drop events in all three cases. For the rectangular and spherical indenters, the load-displacement curves have a linear dependence, which means a linear elastic regime in indentation process, but the elastic stage produced by the spherical indenter does not last long than that produced by the rectangular indenter. For a Berkovich indenter, there is almost no linear elastic regime, the load drop events tend to be small or not noticeable.

2. For the rectangular and spherical indenters, the stress distributions at low indentation depths are smooth with no sudden irregularities, indicating a fully elastic response. At larger indentation depths, the stress distribution becomes more and more irregular with locally very high stresses. For the Berkovich indenter, a sudden increase in stress near the indenter, and the other parts of the specimen is very low, which is the reason for the stress concentration of the sharp indenter. These irregular stress distributions and the formations of pop-in occurred are closely related to dislocation activities during nanoindentation.

3. In all three cases, prismatic dislocation loops are observed to nucleate on the {111} planes, which is in agreement with the predictions of dislocation theory

in a face-centered cubic metal. The dislocation activities are the most complex for the rectangular indenter case, and the next for the spherical indenter case, whereas very few dislocations and an amorphous region directly below the indenter tip are observed in the Berkovich indenter case, which is related to the extremely singular stress field around the indenter tip. According to these results, we can conclude that the shapes of indenters have different and significant influences on dislocation activities and stress distributions during nanoindentation.

Acknowledgements

The work was supported by National Natural Science Foundation of China (Grant Nos. 11772236, 11711530643, and 11472195).

Author details

Wen-Ping Wu[1,2]*, Yun-Li Li[1,2] and Zhennan Zhang[1,2]

1 Department of Engineering Mechanics, School of Civil Engineering, Wuhan University, Wuhan, China

2 State Key Laboratory of Water Resources and Hydropower Engineering Science, Wuhan University, Wuhan, China

*Address all correspondence to: wpwu@whu.edu.cn

References

[1] Fischer-Cripps AC. Nanoindentation. 3rd ed. New York, USA: Spring-Verlag; 2011

[2] Schuh CA. Nanoindentation studies of materials. Materials Today. 2006;**9**(5):32-40

[3] Kiely JD, Houston JE. Nanomechanical properties of Au (111), (001), (110) surface. Physical Review B. 1998;**57**:12588

[4] Li J, Vliet KJV, Zhu T, Yip S, Suresh S. Atomistic mechanism governing elastic limit and incipient plasticity in crystals. Nature. 2002;**418**:307-310

[5] Lee Y, Park JY, Kim SY, Jun S, Im S. Atomistic simulations of incipient plasticity under Al (111) nanoindentation. Mechanics of Materials. 2005;**37**:1035-1048

[6] Liang HY, Woo CH, Huang H, Ngan AHW, Yu TX. Dislocation nucleation in the initial stage during nanoindentation. Philosophical Magazine. 2003;**83**:3609-3622

[7] Lilleodden ET, Zimmerman JA, Foiles SM, Nix WD. Atomistic simulations of elastic deformation and dislocation nucleation during nanoindentation. Journal of the Mechanics and Physics of Solids. 2003; **51**(5):901-920

[8] Lu Z, Chernatynskiy A, Noordhoek MJ, Sinnott SB, Phillpot SR. Nanoindentation of Zr by molecular dynamics simulation. Journal of Nuclear Materials. 2015;**467**: 742-757

[9] DaSilva CJ, Rino JP. Atomistic simulation of the deformation mechanism during nanoindentation of gamma titanium aluminide. Computational Materials Science. 2012;**62**:1-5

[10] Chan CY, Chen YY, Chang SW, Chen CS. Atomistic studies of nanohardness size effects. International Journal of Theoretical and Applied Multiscale Mechanics. 2011;**2**(1):62-71

[11] Imran M, Hussain F, Rashid M, Ahmad SA. Dynamic characteristics of nanoindentation in Ni: A molecular dynamics simulation study. Chinese Physics B. 2012;**21**(11):116201

[12] Noreyan A, Amar JG, Marinescu I. Molecular dynamics simulations of nanoindentation of β-SiC with diamond indenter. Materials Science and Engineering B. 2005;**117**:235-240

[13] Yaghoobi M, Voyiadjis GZ. Effect of boundary conditions on the MD simulation of nanoindentation. Computational Materials Science. 2014; **95**:626-636

[14] Kim KJ, Yoon JH, Cho MH, Jang H. Molecular dynamics simulation of dislocation behavior during nanoindentation on a bicrystal with a Σ=5 (210) grain boundary. Materials Letters. 2006;**60**:3367-3372

[15] Fu T, Peng X, Chen X, Wen S, Hu N, Li Q , et al. Molecular dynamics simulation of nanoindentation on Cu/ Ni nanotwinned multilayer films using a spherical indenter. Scientific Reports. 2016;**6**:35665

[16] Yuan L, Xu Z, Shan D, Guo B. Atomistic simulation of twin boundaries effect on nanoindention of Ag (111) films. Applied Surface Science. 2012;**258**:6111-6115

[17] Fang TH, Weng CI, Chang JG. Molecular dynamics analysis of temperature effects on nanoindentation measurement. Materials Science and Engineering A. 2003;**357**:7-12

[18] Wu CD, Fang TH, Sung PH, Hsu QC. Critical size, recovery, and

mechanical property of nanoimprinted Ni-Al alloys investigation using molecular dynamics simulation. Computational Materials Science. 2012;**53**:321-328

[19] Hansson P. Influence of the crystallographic orientation and thickness of thin copper coatings during nanoindentation. Engineering Fracture Mechanics. 2015;**150**:143-152

[20] Tsuru T, Shibutani Y. Anisotropic effects in elastic and incipient plastic deformation under (001), (110) and (111) nanoindentation of Al and Cu. Physical Review B. 2007;**75**:035415

[21] Remington TP, Ruestes CJ, Bring EM, Remington BA, Lu CH, Kad B, et al. Plastic deformation in nanoindentation of tantalum: A new mechanism for prismatic loop formation. Acta Materialia. 2014;**78**:378-393

[22] Zhao Y, Peng X, Fu T, Sun R, Feng C, Wang Z. MD simulation of nanoindentation on (001) and (111) surfaces of Ag–Ni multilayers. Physica E. 2015;**74**:481-488

[23] Chamani M, Farrahi GH, Movahhedy MR. Molecular dynamics simulation of nanoindentation of nanocrystalline Al/Ni multilayers. Computational Materials Science. 2016;**112**:175-184

[24] Chocyk D, Zientarski T. Molecular dynamics simulation of Ni thin films on Cu and Au under nanoindentation. Vacuum. 2018;**147**:24-30

[25] Lai CW, Chen CS. Influence of indenter shape on nanoindentation: An atomistic study. Interaction and Multiscale Mechanics. 2013;**6**(3): 301-306

[26] Voyiadjis GZ, Yaghoobi M. Large scale atomistic simulation of size effects during nanoindentation: Dislocation

length and hardness. Materials Science and Engineering A. 2015;**634**:20-31

[27] Verkhovtsev AV, Yakubovich AV, Sushko GB, Hanauske M, Solov'yov AV. Molecular dynamics simulations of the nanoindentation process of titanium crystal. Computational Materials Science. 2013;**76**:20-26

[28] Plimpton SJ. Fast parallel algorithms for short-range molecular dynamics. Journal of Computational Physics. 1995;**117**:1-19. Available at: http://lammps.sandia.gov/

[29] Mishin Y, Farkas D, Mehl MJ, Papaconstantopoulos DA. Interatomic potentials for monatomic metals from experimental data and ab initio calculations. Physical Review B. 1999;**59**:3393-3407

[30] Nair AK, Parker E, Gaudreau P, Farkas D, Kriz RD. Size effects in indentation response of thin films at the nanoscale: A molecular dynamics study. International Journal of Plasticity. 2008;**24**:2016-2031

[31] Karimi M, Roarty T, Kaplan T. Molecular dynamics simulations of crack propagation in Ni with defects. Modelling and Simulation in Materials Science and Engineering. 2006;**14**:1409-1420

[32] Shi Y, Falk ML. Stress-induced structural transformation and shear banding during simulated nano-indentation of a metallic glass. Acta Materialia. 2007;**55**:4317-4324

[33] Honeycutt JD, Andersen HC. Molecular dynamics study of melting and freezing of small Lennard-Jones clusters. The Journal of Physical Chemistry. 1987;**91**:4950-4963

[34] Li J. Atomeye: An efficient atomistic configuration viewer. Modelling and Simulation in Materials Science and Engineering. 2003; **11**:173-177

[35] Born M, Huang K. Dynamical Theory of Crystal Lattices. Clarendon Press, Oxford; 1954

[36] Horstemeyer MF, Baskes MI. Atomistic finite deformation simulations: A discussion on length scale effects in relation to mechanical stresses. Journal of Engineering Materials and Technology. 1999; **121**:114-119

10

Physics of Absorption and Generation of Electromagnetic Radiation

Sukhmander Singh, Ashish Tyagi and Bhavna Vidhani

Abstract

The chapter is divided into two parts. In the first part, the chapter discusses the theory of propagation of electromagnetic waves in different media with the help of Maxwell's equations of electromagnetic fields. The electromagnetic waves with low frequency are suitable for the communication in sea water and are illustrated with numerical examples. The underwater communication have been used for the oil (gas) field monitoring, underwater vehicles, coastline protection, oceanographic data collection, etc. The mathematical expression of penetration depth of electromagnetic waves is derived. The significance of penetration depth (skin depth) and loss angle are clarified with numerical examples. The interaction of electromagnetic waves with human tissue is also discussed. When an electric field is applied to a dielectric, the material takes a finite amount of time to polarize. The imaginary part of the permittivity is corresponds to the absorption length of radiation inside biological tissue. In the second part of the chapter, it has been shown that a high frequency wave can be generated through plasma under the presence of electron beam. The electron beam affects the oscillations of plasma and triggers the instability called as electron beam instability. In this section, we use magnetohydrodynamics theory to obtain the modified dispersion relation under the presence of electron beam with the help of the Poisson's equation. The high frequency instability in plasma grow with the magnetic field, wave length, collision frequency and the beam density. The growth rate linearly increases with collision frequency of electrons but it is decreases with the drift velocity of electrons. The real frequency of the instability increases with magnetic field, azimuthal wave number and beam density. The real frequency is almost independent with the collision frequency of the electrons.

Keywords: electromagnetic waves, permittivity, skin depth, loss angle, absorption, Dispersion equations, electron collisions, growth rate, Hall thruster, beam, resistive instability

1. Introduction

X-rays are used to detect bone fracture and determine the crystals structure. The electromagnetic radiation are also used to guide airplanes and missile systems. Gamma rays are used in radio therapy for the treatment of cancer and tumor Gamma rays are used to produce nuclear reaction. The earth get heat from Infrared waves. It is used to kill microorganism. Ultraviolet rays are used for the sterilizing of

The Electromagnetic Spectrum

Frequency (Hz)	Nature	Wavelength (m)	Production	Applications
10^{22}	gamma rays	10^{-13}	Nuclear decay	Cosmic rays
10^{21}	gamma rays	10^{-12}	Nuclear decay	Cancer therapy
10^{18}	x rays	10^{-9}	Inner electronic transitions and fast collisions	Medical diagnosis
10^{16}	ultraviolet	10^{-7}		Sterilization
10^{15}	visible	10^{-6}	Thermal agitation and electronic transitions	Vision, astronomy, optical
6.5×10^{14}	blue	4.6×10^{-7}		
5.6×10^{14}	green	5.4×10^{-7}		
3.9×10^{14}	red	7.6×10^{-7}		
10^{14}	infrared	10^{-5}	Thermal agitation and electronic transitions	Heating, night vision, optical communications
10^{9}	UHF	10^{-3}	Accelerating charges and thermal agitation	Microwave ovens
10^{10}	EHF	10^{-1}		Remote sensing
10^{8}	TV FM	10		Radio transmission
10^{6}	AM	10^{3}		Radio signals
10^{4}	RF	10^{5}	Accelerating charges	

Table 1.
The electromagnetic spectrum.

surgical instruments. It is also used for study molecular structure and in high resolving power microscope. The color of an object is due to the reflection or transmission of different colors of light. For example, a fire truck appears red because it reflects red light and absorbs more green and blue wavelengths. Electro-magnetic waves have a huge range of applications in broadcasting, WiFi, cooking, vision, medical imaging, and treating cancer. Sequential arrangement of electro-magnetic waves according to their frequencies or wave lengths in the form of distinct of groups having different properties in called electromagnetic spectrum. In this section, we discuss how electromagnetic waves are classified into categories such as radio, infrared, ultraviolet, which are classified in **Table 1**. We also sum-marize some of the main applications for each range of electromagnetic waves. Radio waves are commonly used for audio communications with wavelengths greater than about 0.1 m. Radio waves are produced from an alternating current flowing in an antenna.

2. Current status of the research

Underwater communications have been performed by acoustic and optical systems. But the performance of underwater communications is affected by multipath propagation in the shallow water. The optical systems have higher propagation speed than underwater acoustic waves but the strong backscattering due to suspended particles in water always limits the performance of optical systems [1]. UV radiation, free radicals and shock waves generated from electromagnetic fields are effectively used to sterilize bacteria. Pulsed electromagnetic fields (streamer discharge) in water are employed for the sterilization of bacteria. For biological applications of pulsed electromagnetic field, electroporation is usually used to ster-ilize bacteria. This technique is commonly applied for sterilization in food processing. The cells in the region of tissue hit by the laser beam (high intensities $\sim$10 – 100 W/cm^2) usually dies and the resulting region of tissue burn is called a photocoagulation burn. Photocoagulation burns are used to destroy tumors, treat eye conditions and stop bleeding.

Electromagnetic waves in the RF range can also be used for underwater wireless communication systems. The velocity of EM waves in water is more than 4 orders faster than acoustic waves so the channel latency is greatly reduced. In addition, EM waves are less sensitive than acoustic waves to reflection and refraction effects in shallow water. Moreover, suspended particles have very little impact on EM waves. Few underwater communication systems (based on EM waves) have been proposed in reference [2, 3]. The primary limitation of EM wave propagation in water is the high attenuation due to the conductivity of water. For example, it has been shown in [4] that conventional RF propagation works poorly in seawater due to the losses caused by the high conductivity of seawater (typically, 4 S/m). However, fresh water has a typical conductivity of only 0.01 S/m, which is 400 times less than the typical conductivity of seawater. Therefore, EM wave propagation can be more efficient in fresh water than in seawater. Jiang, and Georgakopoulos analyzed the propagation and transmission losses for a plane wave propagating from air to water (frequency range of 23 kHz to 1 GHz). It has been depicted that the propagation loss increases as the depth increases, whereas the transmission loss remains the same for all propagation depths [5]. Mazharimousavi et al. considered variable permeability and permittivity to solve the wave equation in material layers [6]. The Compton and Raman scattering effects are widely employed in the concept of free electron lasers. These nonlinear effects have great importance for fusion physics, laser-plasma acceleration and EM-field harmonic generation. Matsko and Rostovtsev investigated the behavior of overdense plasmas in the presence of the Electromag-netic fields, which can lead to the nonlinear effects such as Raman scattering, modulational instability and self-focusing [7]. The increasing relativistic mass of the particles can make plasma transparent in the presence of high intense the electro-magnetic field change the properties of plasmas [8]. The models of electromagnetic field generated in a non absorbing anisotropic multilayer used to study the optical properties of liquid crystals and propagation of electromagnetic waves in magneto active plasmas [9]. Pulse power generator based on electromagnetic theory has applications such as water treatment, ozone generation, food processing, exhaust gas treatment, engine ignition, medical treatment and ion implantation. The similar work was reviewed by Akiyama et al. [10]

Applications for environmental fields involving the decomposition of harmful gases, generation of ozone, and water treatment by discharge plasmas in water utilizing pulsed power discharges have been studied [11–14]. High power micro-wave can be involved to joining of solid materials, to heat a surface of dielectric material and synthesis of nanocomposite powders. Bruce et al. used a high-power millimeter wave beam for joining ceramics tubes with the help of 83-GHz Gyrotron [15]. The use of shock waves to break up urinary calculi without surgery, is called as extracorporeal shock wave lithotripsy. Biofilm removal to inactivation of fungi, gene therapy and oncology are the interesting uses of shock waves lithotripsy. Loske overviewed the biomedical applications (orthopedics, cardiology, traumatology, rehabilitation, esthetic therapy) of shock waves including some current research. [16]. Watts et al. have reported the theory, characterization and fabrications of metamaterial perfect absorbers (MPAs) of electromagneti c waves. The motivation for studying MPAs comes mainly from their use in potential applications as selective thermal emitters in automotive radar, in local area wireless network at the frequency range of 92–95 GHz and in imaging at frequency 95 and 110 GHz. [17]. Ayala investigated the applications of millimeter waves for radar sensors. [18] Metamaterial perfect absorbers are useful for spectroscopy and imaging, actively integrated photonic circuits and microwave-to-infrared signature control [19–21]. In [22, 23], authors show the importance of THz pulse imaging system for characterizing biological tissues such as skin, muscle and veins. Reference [24] reported the propagation of EM waves on a graphene sheet. The Reference [25] compared the CNT-based nano dipole antenna and GNR-based nano patch antenna. Due to short wavelength, even a minute variations in water contents and biomaterial tissues can be detected by terahertz radiations due to existence of molecular resonances at such frequencies. Consequently, one of the emerging areas of research is analyzing the propagation of terahertz electromagnetic waves through the tissues to develop diagnostic tools for early detection and treatment such as abnormalities in skin tissues as a sign of skin cancer [26]. Shock waves may stimulate osteogenesis and chondrogenesis effects [27], induce analgesic effects [28] and tissue repair mecha-nisms [29]. Shock waves therapy are also used to treat oncological diseases and other hereditary disorders [27, 30]. Chen et al. proposed a mathematical model for the propagating of electromagnetic waves coupling for deep implants and simulated through COMSOL Multiphysics [31]. Body area networks technological is used to monitor medical sensors implanted or worn on the body, which measure important physical and physiological parameters [32, 33]. Marani and Perri reviewed the aspects of Radio Frequency Identification technology for the realization of minia-turized devices, which are implantable in the human body [34]. Ultrasonic can transport high power and can penetrate to a deeper tissue with better power effi-ciency. [35, 36]. Ref. [37], discuss the radar-based techniques to detect human motions, wireless implantable devices and the characterization of biological mate-rials. Low frequency can deliver more power with deeper penetrating ability in tissue [38, 39]. Contactless imaging techniques based on electromagnetic waves are under continuous research. Magnetic resonance imaging technology and physiolog-ical processes of biological tissues and organisms [40, 41]. The electrical properties of biological mediums are found very useful because it is related to the pathological and physiological state of the tissues [42–44].

3. Interaction of electromagnetic wave fields with biological tissues

From last decade, researchers are interested about biological effects of electromagnetic energy due to public concern with radiation safety and measures. The electromagnetic energy produces heating effects in the biological tissues by increasing the kinetic energy of the absorbing molecules. Therefore the body tissues absorb strongly in the UV and in the Blue/green portion of the spectrum and transmit reds and IR. A surgeon can select a particular laser to target cells for photovaporization by determining which wavelengths your damaged cell will absorb and what the surrounding tissue will not. The heating of biological tissues depends on dielectric properties of the tissues, tissue geometry and frequency of the source. The tissues of the human body are extremely complex. Biological tissues are composed of the extracellular matrix (ECM), cells and the signaling systems. The signaling systems are encoded by genes in the nuclei of the cells. The cells in the tissues reside in a complex extracellular matrix environment of proteins, carbohydrates and intracellular fluid composed of several salt ions, polar water molecules and polar protein mo lecules. The dielectric constant o f tissues decreases as the frequency is increased to GHz level. The effective conductivity, rises with frequency. The tissues of brain, muscle, liver, kidney and heart have larger dielectric constant and conductivity as compared to tissues of fat, bone and lung. The action of electromagnetic fields on the tissues produce the rotation of dipole molecules at the frequency of the applied electromagnetic energy which in turn affects the displacement current through the medium with an associated dielectric loss due to viscosity. The electromagnetic field also produce the oscillation of the free charges, which in turn gives rise to conduction currents with an associated energy loss due to electrical resistance of the medium. The interaction of electromagnetic wave fields with biological tissues is related to dielectric properties. Johnson and Guy reviewed the absorption and scattering effects of light in biological tissues [45]. In ref. [46], the method of warming of human blood from refrigerated (bank blood storage temperature $\sim$ 4 to 6°C) has been discussed with the help of microwave.

4. Complex dielectric permittivity

The dielectric permittivity of a material is a complex number containing both real and imaginary components. It describes a material's ability to permit an electric field. It dependent on the frequency, temperature and the properties of the mate-rial. This can be expressed by

$$\varepsilon_c = \varepsilon_0(\varepsilon' - j\varepsilon'') \tag{1}$$

where ε' is the dielectric constant of the medium. The ε'' is called the loss factor of the medium and related with the effective conductivity such that $\varepsilon'' = \frac{\sigma}{\varepsilon_0 \omega}$. These coefficients are related through by loss tangent $\tan\delta = \frac{\varepsilon''}{\varepsilon'}$. In other words loss factor is the product of loss tangent and dielectric constant, that is $\varepsilon'' = \varepsilon' \tan\delta$. The loss tangent depends on frequency, moisture content and temperature. If all energy is dissipated and there is no charging current then the loss tangent would tend to infinity and if no energy is dissipated, the loss tangent is zero [45, 47–49] The high power electromagnetic waves are used to generate plasma through laser plasma interaction. Gaseous particles are ionized to bring it in the form of plasma through injection of high frequency microwaves. The electrical permittivity in plasma is affected by the plasma density [50]. If the microwave electric field ($\tilde{E}$) and the velocity ($\tilde{v}$) are assumed to be varying with $e^{i\omega t}$, the plasma dielectric constant can be read as,

$$\epsilon = \epsilon_0 \left(1 - \frac{\omega_{pe}^2}{\omega^2} \right) \tag{2}$$

Where; the ω_{pe} is the electron plasma frequency and given by the relation,

$$\omega_{pe} = \sqrt{\frac{n_e e^2}{\epsilon_0 m_e}} \tag{3}$$

Recently many researchers have studied the plasma instabilities in a crossed field devices called Hall thrusters (space propulsion technology). The dispersion rela-tions for the low and high frequency electrostatic and electromagnetic waves are derived in the magnetized plasma. The dispersion relations for the resistive and Rayliegh Taylor instabilities has been derived for the propagation of waves in a magnetized plasma under the effects of various parameters [51–61].

5. Propagation of EM fields (waves) in conductors

The behavior of EM waves in a conductor is quite different from that in a source-free medium. The conduction current in a conductor is the cause of the difference. We shall analyze the source terms in the Maxwell's equations to simplify Maxwell's equations in a conductor. From this set of equations, we can derive a diffusion equation and investigate the skin effects.

5.1 Gauss' law for electric field

The Electric flux φ_E through a closed surface A is proportional to the net charge q enclosed within that surface.

$$\varphi_E = \oint \vec{E} \cdot \hat{n} dA = \frac{q}{\varepsilon_0} = \frac{1}{\varepsilon_0} \int_V \rho dV \qquad (4)$$

$$\text{Differential form, } \vec{\nabla} \cdot \vec{E} = \frac{\rho}{\varepsilon_0} \qquad (5)$$

5.2 Faraday's law

The electromagnetic force induced in a closed loop is proportional to the negative of the rate of change of the magnetic flux, φ_B through the closed loop,

$$\oint \vec{E} \cdot d\vec{l} = \frac{\partial \varphi_B}{\partial t} = \frac{\partial}{\partial t} \oint \vec{B} \cdot dA \qquad (6)$$

Faraday's law in differential form,

$$\vec{\nabla} \times \vec{E} = -\frac{\partial \vec{B}}{\partial t} \qquad (7)$$

5.2 Magnetic Gauss's law for magnetic field

The Magnetic flux φ_B through a closed surface, A is equal to zero.

$$\varphi_B = \oint \vec{B} \cdot dA = 0 \qquad (8)$$

In the differential form

$$\vec{\nabla} \cdot \vec{B} = 0 \qquad (9)$$

5.3 Ampere's law

The path integral of the magnetic field around any closed loop, is proportional to the current enclosed by the loop plus the displacement current enclosed by the loop.

$$\oint \vec{B} \cdot d\vec{l} = \mu_0 I + \mu_0 \varepsilon_0 \frac{\partial \varphi_E}{\partial t} \qquad (10)$$

Ampere's law in differential form

$$\vec{\nabla} \times \vec{B} = \mu \sigma \vec{E} + \mu \varepsilon \frac{\partial \vec{E}}{\partial t} \qquad (11)$$

6. Properties of plane wave (monochromatic) in vacuum

Let us assume that the wave equations (fields) has the solution in the form of $\vec{E}\left(\vec{B}\right) = \vec{E}_0\left(\vec{B}_0\right)e^{-i(kz-\omega t)}$, then the vector operators can be written as $\nabla \rightarrow -ik$ and $\frac{\partial}{\partial t} \rightarrow i\omega$.

a. The vector k and fields $\vec{E}\left(\vec{B}\right)$ are perpendicular

From Gauss's law $k \cdot E = 0$

b. The field $\vec{B}$ is perpendicular to the vector k and field $\vec{E}$

From Faraday's law $-i\vec{k} \times \vec{E} = -i\omega\vec{B}$

$$\Rightarrow \vec{B} = \frac{\vec{k} \times \vec{E}}{\omega} = \frac{k\hat{k} \times \vec{E}}{\omega} = \frac{\hat{k} \times \vec{E}}{c} \tag{12}$$

Where we have used $\omega = ck$ and unit vector $\hat{k} = \vec{k}/k$. This implies that all three vectors are perpendicular to one another (**Figure 1**).

Let us apply curl operator to the 2nd equation.
Maxwell's equation:

$$\vec{\nabla} \times \left(\vec{\nabla} \times \vec{E}\right) = -\vec{\nabla} \times \left(\frac{\partial \vec{B}}{\partial t}\right) = -\frac{\partial}{\partial t}\left(\vec{\nabla} \times \vec{B}\right) = -\frac{\partial}{\partial t}\left(\mu\sigma\vec{E} + \mu\varepsilon\frac{\partial \vec{E}}{\partial t}\right) \tag{13}$$

So

$$-\nabla^2\vec{E} = -\mu\sigma\frac{\partial \vec{E}}{\partial t} - \mu\varepsilon\frac{\partial^2 \vec{E}}{\partial t^2} \tag{14}$$

Similar, the magnetic field satisfy the same equation

$$-\nabla^2\vec{B} = -\mu\sigma\frac{\partial \vec{B}}{\partial t} - \mu\varepsilon\frac{\partial^2 \vec{B}}{\partial t^2} \tag{15}$$

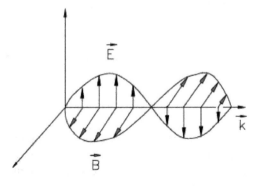

Figure 1.
Orientations of electric field, magnetic field and wave vector.

6.1 Skin depth

Suppose we have a plane wave field. It comes from the $-z$ direction and reaches a large conductor. Surface at $z = 0$ outside of a conductor: $E = E_0 e^{-i\omega t} e_x$ at $z = 0$. Looking for the wave like solution of electric (magnetic) fields by assuming the wave inside the conductor has the form, where k is an unknown constant. Suppose, the waves are traveling only in the z direction (no x or y components). These waves are called plane waves, because the fields are uniform over every plane perpendicular to the direction of propagation. We are interested, then, in fields of the form

$$\vec{E}\left(\vec{B}\right) = \vec{E}_0\left(\vec{B}_0\right)e^{-i(kz - \omega t)} \tag{16}$$

for the waves of the above type, we find from the diffusion equation

$$k^2 E = -i\omega\mu\sigma E + \mu\varepsilon\omega^2 E \tag{17}$$

Or $(k^2 + i\omega\mu\sigma - \mu\varepsilon\omega^2)E = 0$

$$\text{For non}-\text{trivial solution } k^2 + i\omega\mu\sigma - \mu\varepsilon\omega^2 = 0 \tag{18}$$

The presence of imaginary term due to conductivity of the medium gives different dispersion relation from the dielectric medium. From Eq. (18) we can expect the wave vector to have complex form.

Let us write

$$\vec{k} = \vec{\alpha} - i\vec{\beta} \tag{19}$$

Here the real part $\vec{\alpha}$ determine the wavelength, refractive index and the phase velocity of the wave in a conductor. The imaginary part $\vec{\beta}$ corresponds to the skin depth in a conductor. The solutions of Eqs. (18) and (19), gives the real and imaginary part of wave vector k in terms of materials' properties.

$$\alpha = \omega\sqrt{\frac{\varepsilon\mu}{2}}\left[\sqrt{1 + \frac{\sigma^2}{\varepsilon^2\omega^2}} + 1\right]^{\frac{1}{2}} \tag{20}$$

$$\text{And } \beta = \frac{\omega\mu\sigma}{2\alpha} \tag{21}$$

$$\text{Or } \beta = \omega\sqrt{\frac{\varepsilon\mu}{2}}\left[\sqrt{1 + \frac{\sigma^2}{\varepsilon^2\omega^2}} - 1\right]^{\frac{1}{2}} \tag{22}$$

If we use complex wave vector $\vec{k} = \vec{\alpha} - i\vec{\beta}$ into Eq. (16), then the wave equation for a conducting medium can be written as

$$\vec{E} = \vec{E}_0 e^{-\beta z}e^{-i(\alpha z - \omega t)} \tag{23}$$

It is clear from the above equation that the conductivity of the medium affects the wavelength for a fixed frequency. The first exponential factor $e^{-\beta z}$ gives an exponential decay in the amplitude (with increasing z) of the wave as shown in **Figure 2**. The cause of the decay of the amplitude of the wave can be explained in a very precise way in terms of conservation of energy. Whenever the incoming

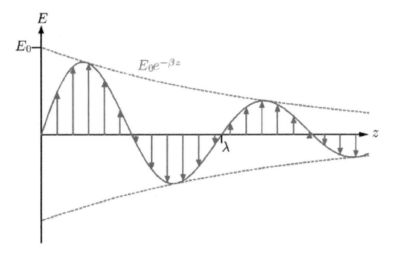

Figure 2.
Decaying of electromagnetic field.

electromagnetic radiation interacts with a conducting material, it produces current in the in the conductor. The current produces Joule heating effect which must be compensated from the energy of the wave. Therefore we can expect the decay in the amplitude of the wave. The second factor $e^{-i(\alpha z - \omega t)}$ gives the plane wave variations with space and time.

7. Alternating magnetic field in a conducting media

From Faraday's law, the both fields are related by

$$\vec{k} \times \vec{E}_0 = \omega \vec{B}_0 \tag{24}$$

$$\text{Or } \vec{B}_0 = \frac{\vec{k} \times \vec{E}_0}{\omega} \tag{25}$$

Thus as in dielectric case, both fields are perpendicular to each other and also perpendicular to the direction of motion with same phase angle.

7.1 Phase change in fields in a conducting media

The complex wave vector k, gives the phase angle between the fields in a conducting medium. Let us assume that E is polarized along the x direction

$$\vec{E} = \hat{i}\vec{E}_0 e^{-\beta z} e^{-i(\alpha z - \omega t)} \tag{26}$$

And the magnetic field results from Eq. (25) is given by

$$\vec{B} = \hat{j}\frac{|k|}{\omega}\vec{E}_0 e^{-\beta z} e^{-i(\alpha z - \omega t)} \tag{27}$$

From Eq. (19), the complex number k can be written as

$$|k| = \sqrt{\alpha^2 + \beta^2}e^{i\varphi} = \text{R}e^{i\varphi} \tag{28}$$

$$\text{Thus } R = \sqrt{\alpha^2 + \beta^2} = \omega\sqrt{\varepsilon\mu}\left[\sqrt{1 + \frac{\sigma^2}{\varepsilon^2\omega^2}}\right]^{\frac{1}{4}} \tag{29}$$

$$\text{And the phase angle } \varphi = \tan^{-1}\frac{\beta}{\alpha} \tag{30}$$

Further if the initial phases of the fields are φ_E and φ_B, then the amplitude are given by

$$\vec{E}_0 = E_0 e^{i\varphi_E} \tag{31}$$

$$\vec{B}_0 = B_0 e^{i\varphi_B} \tag{32}$$

From Eq. (27)

$$B_0 e^{i\varphi_B} = \frac{\text{R}e^{i\varphi}}{\omega} E_0 e^{i\varphi_E} \tag{33}$$

Therefore, the both fields are out of phase with angle

$$\varphi = \varphi_B - \varphi_E \tag{34}$$

From Eq. (33) the ratio of the magnetic field to the electric field is

$$\frac{B_0}{E_0} = \frac{R}{\omega} \tag{35}$$

By using Eq. (29)

$$\frac{B_0}{E_0} = \sqrt{\varepsilon\mu}\left[1 + \frac{\sigma^2}{\varepsilon^2\omega^2}\right]^{\frac{1}{4}} \tag{36}$$

In other words, we can say that the magnetic field advanced from electric field by the phase angle φ. In terms of sinusoidal form, these fields follow the following expressions.

$$\vec{E} = \hat{i}\vec{E}_0 e^{-\beta z}\cos(\omega t - \alpha z + \varphi_E) \tag{37}$$

And the magnetic field results from Eq. (32) is given by

$$\vec{B} = \hat{j}\vec{B}_0 e^{-\beta z}\cos(\omega t - \alpha z + \varphi_E + \varphi) \tag{38}$$

The above equations direct that the amplitude of an electromagnetic wave propagating (through a conductor) decays exponentially on a characteristic length scale, d, that is known as the skin-depth [48].

7.2 Skin depth

Skin depth measure the distance that the wave travels before it's amplitude falls to 1/e of its original value [48]. From Eq. (37), the amplitude of the wave falls by a factor $1/e$ in a distance $z = \frac{1}{\beta}$. In other words it is a measure of how far the wave penetrates into the conductor. Mathematically skin depth is denoted by δ, therefore

$$\delta = \frac{1}{\beta} \tag{39}$$

If we study poor conductor, which satisfies the inequality $\sigma << \varepsilon\omega$, then Eqs. (20) and (21) leads to

$$\alpha \approx \omega\sqrt{\varepsilon\mu} \tag{40}$$

And we know that $\beta = \dfrac{\omega\mu\sigma}{2\alpha} \tag{41}$

Substitute the value of α into Eq. (32), we get

$$\text{Or } \beta \approx \omega\frac{\mu\sigma}{2\omega\sqrt{\varepsilon\mu}} \tag{42}$$

$$\text{Or } \beta \approx \frac{\sigma}{2}\sqrt{\frac{\mu}{\varepsilon}} \tag{43}$$

The phase velocity $V_{ph} = \dfrac{\omega}{\alpha} \approx \dfrac{1}{\sqrt{\varepsilon\mu}} \tag{44}$

Skin depth in poor conductor $\delta = \dfrac{1}{\beta} = \dfrac{2}{\sigma}\sqrt{\dfrac{\varepsilon}{\mu}} \tag{45}$

So, it is independent from the frequency.

The Eq. (41) state that at higher frequency, the absorbing parameter lost its significance that is $\beta << \alpha$. We can conclude that at higher frequency the wavelength does not decay very fast in a poor conductor. Moreover both the fields are also in same phase by the relation $\omega B_0 = \alpha E_0$. Also the phase velocity is independent from the frequency [47].

8. Wave propagation in perfect conductors

The transmission lines and communication systems are made up with silver, copper and aluminum. In most cases these conductors satisfies the inequality $\sigma >> \varepsilon\omega$, then Eqs. (20) and (21) leads to

$$\alpha \approx \sqrt{\frac{\omega\mu\sigma}{2}} \tag{46}$$

$$\text{And } \beta \approx \sqrt{\frac{\omega\mu\sigma}{2}} \tag{47}$$

$$\text{Therefore } \beta \approx \alpha \tag{48}$$

$$\text{Skin depth } \delta = \frac{1}{\beta} = \frac{1}{\alpha} = \sqrt{\frac{2}{\omega\mu\sigma}} \tag{49}$$

The wave decays significantly within one wavelength. Since $\delta \propto \sqrt{1/\omega\sigma}$, the deep penetration occurs, when the inequality $\sigma << \varepsilon\omega$ is satisfied (at Low frequency in a Poor conductor).

9. Electromagnetic wave propagation into water

EM wave propagation can be more efficient in fresh water than in seawater. The radiofrequency wave propagation works poorly in seawater due to the losses caused by the high conductivity of seawater. The limitation of EM wave propagation in water is the high attenuation due to the conductivity of water (typically, 4 S/m), however fresh water has a conductivity of 0.01 S/m. These properties are used to construct underwater sensor network based on electromagnetic waves to trace out the natural resources buried underwater, where the conventional optical water sensors are difficult to utilize in an underwater environment due to backscatter and absorptions [47].

Example:

For sea water,

$\mu = \mu_0 = 4\pi \times 10^{-7} N/A^2, \varepsilon \cong 81\varepsilon_0$ and $\sigma \approx 5(\Omega.m)^{-1}$.

The skin depth in poor conductor

$$\delta = \frac{2}{\sigma}\sqrt{\frac{\varepsilon}{\mu}} = \frac{2}{\sigma}\sqrt{\frac{81\varepsilon_0}{\mu_0}} \tag{50}$$

$$= \frac{2\sqrt{81}}{\sigma Z} = \frac{18}{5 \times 377} \approx 0.96 \text{cm}. \tag{51}$$

If the sea water satisfies the inequality $\sigma << \varepsilon\omega$, of poor conductor, which require

$$f = \frac{\omega}{2\pi} >> \frac{\sigma}{2\pi\varepsilon} = 10^9 Hz \tag{52}$$

Therefore at $10^9 Hz$ or $\lambda << 30 \ cm$, sea water behave as poor conductor. On the other hand at the radio frequency range $f << 10^9$ Hz, the inequality $\sigma >> \varepsilon\omega$, can be satisfied, the skin depth $\delta = \sqrt{2/(\omega\mu\sigma)}$ is quite short. To reach a depth δ = 10 m, for communication with submarines,

$$f = \frac{\omega}{2\pi} = \frac{1}{\pi\sigma\mu\delta^2} \approx 500 Hz \tag{53}$$

The wavelength in the air is about

$$\lambda = \frac{c}{f} = \frac{3 \times 10^8}{500} = 600 \text{km} \tag{54}$$

The skin depth at different frequency in sea water are 277 m at 1 Hz, 8.76 m at 1KHz, 0.277 m at 1 MHz and 0.015 at 1GHz if the conductivity of sea-water is taken to about $\sigma = 3/\Omega m$ and $\varepsilon_r = 80$. These effects leads to severe restrictions for radio communication with submerged submarines. To overcome this, the communication must be performed with extremely low frequency waves generated by gigantic antennas [47].

9.1 Short wave communications

At 60 km to 100 km height from the earth, ionosphere plasma has a typical density of $10^{13}/m^3$, which gives the plasma frequency of order 28 MHz. the waves below this frequency shows reflections from the layer of ionosphere to reach the receiver's end. The conductivity of the earth is 10^{-2} S/m, Earth behave as a good conductor, if the inequality $\sigma > > \varepsilon\omega$ is satisfied. In other word

$$f < < \frac{\sigma}{2\pi\varepsilon} = 180 \tag{55}$$

MHz, therefore below 20 MHz, the earth is good conductor.
Example: skin depth at $f = 60$ Hz for copper.

$$\delta = \sqrt{\frac{2}{2\pi \times 60 \times 4\pi \times 10^{-7} \times 6 \times 10^7}} = 8\text{mm} \tag{56}$$

The frequency dependent skin-depth in Copper ($\sigma = 6\,25 \times 10^7/\Omega m$) can be expressed as $d = \frac{6}{\sqrt{f(\text{Hz})}}$ cm. It says that the skin-depth is about 6 cm at 1 Hz and it reduced to 2 mm at 1 kHz. In other words it conclude that an oscillating electromagnetic signal of high frequency, transmits along the surface of the wire or on narrow layer of thickness of the order the skin-depth in a conductor. In the visible region ($\omega \sim 10^{15}$/s) of the spectrum, the skin depth for metals is on the order of $10A^0$. The skin depth is related with wavelength λ (inside conductor) as

$$\lambda = \frac{2\pi}{\alpha} = 2\pi\sqrt{\frac{2}{\omega\mu\sigma}} \tag{57}$$

The phase velocity $V_{ph} = \frac{\omega}{\alpha} = \frac{\omega\lambda}{2\pi} \approx \sqrt{\frac{2\omega}{\mu\sigma}} \tag{58}$

Therefore for a very good conductor, the real and imaginary part of the wave vector attain the same values. In this case the amplitude of the wave decays very fast with frequency as compared to bad conductor. The phase velocity of the wave in a good conductor depends on the frequency of the electromagnetic light. Consequently, an electromagnetic wave cannot penetrate more than a few skin-depths into a conducting medium. The skin-depth is smaller at higher frequencies. This implies that high frequency waves penetrate a shorter distance into a conductor than low frequency waves.

Question: Find the skin depths for silver at a frequency of 10^{10} Hz.

$$\text{Skin depth } \delta = \sqrt{\frac{2}{\omega\mu\sigma}} \tag{59}$$

$$\delta = \sqrt{\frac{2}{2\pi \times 10^{10} \times 4\pi \times 10^{-7} \times 6.25 \times 10^7}} = 6.4 \times 10^{-4}\text{mm} \tag{60}$$

Therefore, in microwave experiment, the field do not penetrate much beyond .00064 mm, so no point it's coating making further thicker. There is no advantage to construct AC transmission lines using wires with a radius much larger than the skin depth because the current flows mainly in the outer part of the conductor.

Question: wavelength and propagation speed in copper for radio waves at 1 MHz. compare the corresponding values in air (or vacuum). $\mu_0 = 4\pi \times 10^{-7}$ H/m.

From Eq. (40),

$$\lambda_{Cu} = \frac{2\pi}{\alpha_{Cu}} \text{ and } \alpha_{Cu} \approx \sqrt{\frac{\omega\mu\sigma_{Cu}}{2}} \qquad (61)$$

Therefore, $\lambda_{Cu} = 2\pi\sqrt{\frac{2}{\omega\mu\sigma_{Cu}}}$

$$\lambda_{Cu} = 2\pi\sqrt{\frac{2}{2\pi \times 10^6 \times 4\pi \times 10^{-7} \times 6.25 \times 10^7}} = 0.4\text{mm} \qquad (62)$$

The propagating velocity in copper $V_{ph} = \frac{\omega}{\alpha} = \frac{\omega\lambda}{2\pi}$

$$V_{ph} = 0.4 \times 10^{-3} \times 10^6 = 400\text{m/s} \qquad (63)$$

The above parameters are quite different in vacuum as follow

$$\lambda_{Vacuum} = \frac{c}{\nu} = \frac{3 \times 10^8}{10^6} = 300\text{m} \qquad (64)$$

There is no advantage to construct AC transmission lines using wires with a radius much larger than the skin depth because the current flows mainly in the outer part of the conductor.

10. Complex permittivity of bread dough and depth of penetration

After baking for few minutes, the relative permittivity of bread dough at frequency 600 MHz is $\varepsilon_{cr} = 23.1 - j11.85$. Calculate the depth of penetration of microwave.

Solution: the loss tangent of bread dough is

$$\tan\delta = \frac{11.85}{23.1} = 0.513 \qquad (65)$$

The depth of penetration is given as

$$d \approx \frac{c\sqrt{2}}{2\pi f \varepsilon_r' \sqrt{\left(\sqrt{1 + \tan\delta^2} - 1\right)}} \qquad (66)$$

After substituting all the parameters, we get

$$d \approx \frac{\sqrt{2} \times 3 \times 10^8}{2\pi \times 600 \times 10^6} \frac{1}{23.1\sqrt{\left(\sqrt{1 + (0.513)^2} - 1\right)}} \approx 6.65\text{cm} \qquad (67)$$

It is worthy to note that the depth of penetration decreases with frequency.

11. The AC and DC conduction in plasma

Let the collision frequency of electrons with ions and ω the frequency of the EM waves in the conductor. The equation of motion for electrons is:

$$m\frac{dv}{dt} = -eE - m\nu v \tag{68}$$

Assume $v = v_0 e^{-i\omega t}$ and use $\partial/\partial t \rightarrow -i\omega$, we obtain

$$-i\omega m v = -eE - m\nu v \rightarrow v = \frac{-e}{m(\nu - i\omega)}E \tag{69}$$

the current density is expressed by $j = -env$

$$j_f = \frac{-ne^2}{m(\nu - i\omega)}E \tag{70}$$

Therefore, the AC conductivity can be read as

$$\sigma(\omega) = \frac{1}{(\nu - i\omega)}\frac{ne^2}{m} \tag{71}$$

In infrared range $\omega << \nu \sim 10^{14}(1/\sec)$, so the DC conductivity

$$\sigma = \frac{ne^2}{m\nu} \tag{72}$$

can be taken.

Let us now compare the magnitude of conduction current with that of the displacement current.

Assume $E = E_0 e^{-i\omega t}$. Then

$$\left|\frac{j_f}{\varepsilon\frac{\partial E}{\partial t}}\right| = \frac{\sigma E}{\varepsilon\omega E} = \frac{\sigma}{\varepsilon\omega} \tag{73}$$

In copper, $\sigma = 6 \times 10^7 (s/m)$. The condition for $j_f \approx \varepsilon\frac{\partial E}{\partial t}$, or $\frac{\sigma}{\varepsilon\omega} \approx 1$ leads to

$$\omega = \frac{\sigma}{\varepsilon} = \frac{6 \times 10^7}{8.85 \times 10^{-12}} \sim 7 \times 10^{19} (rad/\sec) \tag{74}$$

At frequencies $\omega < 10^{12} (rad/\sec)$ (communication wave frequency), $\frac{\sigma}{\varepsilon\omega} >> 1$ or $\left|j_f\right| >> \left|\varepsilon\frac{\partial E}{\partial t}\right|$.

12. Electromagnetic pulse and high power microwave overview

Several nations and terrorists have a capability to use electromagnetic pulse (EMP) as a weapon to disrupt the critical infrastructures. Electromagnetic pulse is an intense and direct energy field that can interrupt sensitive electrical and electronic equipment over a very wide area, depending on power of the nuclear device and altitude of the burst. An explosion exploded at few heights in the atmosphere can produce EMP and known as high altitude EMP or HEMP. High power microwave (HPM) can be produced with the help of powerful batteries by electrical equipment that transforms battery power into intense microwaves which may be harmful electronics equipments [62–71]. The high- power electromagnetic (HPEM) term describes a set of transient electromagnetic environments with intense electric and magnetic fields. High- power electromagnetic field may be produced by electrostatic discharge, radar system, lightning strikes, etc. The nuclear bursts can lead to the production of electromagnetic pulse which may be used against the enemy country's military satellites. Therefore the sources derived from lasers, nuclear events are vulnerable and called laser and microwave threats. Microwave weapons do not rely on exact knowledge of the enemy system. These weapons can leave persisting and lasting effects in the enemy targets through damage and destruction of electronic circuits, components. Actually HEMP or HPM energy fields, as they instantly spread outward, may also affect nearby hospital equipment or personal medical devices, such as pacemakers. These may damage critical electronic systems throughout other parts of the surrounding civilian infrastructure. HEMP or HPM may damage to petroleum, natural gas infrastructure, transportation systems, food production, communication systems and financial systems [62–71].

13. Generation of high - frequency instability through plasma environment

The beams of ions and electrons are a source of free energy which can be transferred to high power waves. If conditions are favorable, the resonant interaction of the waves in plasma can lead to nonlinear instabilities, in which all the waves grow faster than exponentially and attain enormously large amplitudes. These instabilities are referred to as explosive instabilities. Such instabilities could be of considerable practical interest, as these seem to offer a mechanism for rapid dissi-pation of coherent wave energy into thermal motion, and hence may be effective for plasma heating [72, 73]. A consistent theory of explosive instability shows that in the three-wave approximation the amplitudes of all the waves tend to infinity over a finite time called explosion time [74, 75]. In ref. [74], an explosive- generated –plasma is discovered for low and high frequency instabilities. The solution of dis-persion equation is found numerically for the possibility of wave triplet and syn-chronism conditions. The instabilities is observed to propagate whose wave number.

14. Electron beam plasma model and theoretical calculation

Here we considersions, electrons and negatively charged electron beam are immersed in a Hall thruster plasma channel [51–55]. The magnetic field is consider as $\vec{B} = B\hat{z}$ so that electrons are magnetized while ions remains un-magnetized and electrons rotates with cyclotron frequency $\Omega = \frac{eB}{m_e}$, whereas the gyro-radius for ions is larger so that they cannot rotate and simply ejects out by providing thrust to the device. The axial electric field $\vec{E} = E\hat{x}$ (along the x - axis) which accelerates the particles. It causes electrons have a $\vec{E} \times \vec{B}$ drift in the azimuthal direction (y-axis) whereas the movement of ions is restricted along x-axis. Similar to previous studies, here, we consider the motion of all the species i.e. for ions (density n_i, mass m_i, velocity v_i) for electrons (density n_e, mass m_e, velocity v_e), for electron beam (density n_b, mass m_b, velocity v_b) and collision frequency for the excitation of instability. The basic fluid equations are given as follows:

$$\frac{\partial n_i}{\partial t} + \vec{\nabla} \cdot \left(\vec{v}_i n_i \right) = 0 \tag{75}$$

$$m_i\left(\frac{\partial}{\partial t} + \left(\vec{v}_i.\vec{\nabla}\right)\right)\vec{v}_i = e\vec{E} \tag{76}$$

$$\frac{\partial n_e}{\partial t} + \vec{\nabla}\cdot\left(\vec{v}_e n_e\right) = 0 \tag{77}$$

$$m_e\left(\frac{\partial}{\partial t} + \left(\vec{v}_e\cdot\vec{\nabla}\right) + v\right)\vec{v}_e = -e\left(\vec{E} + \vec{v}_e \times \vec{B}\right) \tag{78}$$

$$\frac{\partial n_b}{\partial t} + \vec{\nabla}\cdot\left(\vec{v}_b n_b\right) = 0 \tag{79}$$

$$m_b\left(\frac{\partial}{\partial t} + \left(\vec{v}_b\cdot\vec{\nabla}\right)\right)\vec{v}_b = -en_b\vec{E} \tag{80}$$

$$\varepsilon_0\nabla^2\varphi_1 = e(n_{e1} - n_{i1} + n_{b1}) \tag{81}$$

Since the larmor radius of ions are larger than the length of the channel (6 cm), therefore ions are considered as unmagnetized in the channel and are accelerated along the axial direction of the chamber. We consider ions initial drift in the positive x – direction ($\vec{v}_{i0} = v_{i0}\hat{x}$) with neglecting motion in both azimuthal and radial directions [51–55]. Electron has motion in the x-direction ($\vec{v}_b = v_b\hat{x}$) since electrons are affected by magnetic field and get magnetized, we takes their $\vec{E} \times \vec{B}$ initial drift in the y – direction ($\vec{v}_e = v_e\hat{y}$).

To find the oscillations by the solutions of the above equations we take the quantities varied as the $A(r,t) = A_0 e^{i(k.r-\omega t)}$ for first order perturb quantities $n_{i1}, n_{e1}, n_{b1}, v_{i1}, v_{e1}, v_{b1}$ and $\vec{E}_1$ together with ω as a frequency of oscillations and the k is the wave propagation vector within plane of (x, y) . On remarking the magnetic fields are large enough in Hall thruster and condition $\Omega >> \omega, k_y v_{e0}, v$ is satisfied [51–56]. By solving the equation of motion and the equation of continuity for electrons, we get the perturbed density of electrons in terms of oscillating potential φ_1 in the following way

$$n_{e1} = \frac{e n_{e0}\hat{\omega}k^2\varphi_1}{m_e\Omega^2\left(\omega - k_y v_{e0}\right)} \tag{82}$$

Let us consider, $\hat{\omega} = \omega - k_y v_{e0} - iv$, the cyclotron frequency $\Omega = \frac{eB}{m_e}$ and $k^2 = k_x^2 + k_y^2$.

Similarly, on solving equation for ions we get the ion density term as

$$n_{i1} = \frac{e k^2 n_{i0}\varphi_1}{m_i(\omega - k_x v_{i0})^2} \tag{83}$$

Similarly for electron beam density given as

$$n_{b1} = -\frac{e k^2 n_{b0}\varphi_1}{m_b(\omega - k_x v_{b0})^2} \tag{84}$$

By putting these density values in the Poisson's equations

$$-k^2\varphi_1 = \frac{e^2 n_{e0}\hat{\omega}k^2\varphi_1}{m_e\varepsilon_0\Omega^2\left(\omega - k_y v_{e0}\right)} - \frac{e^2 k^2 n_{i0}\varphi_1}{m_i\varepsilon_0(\omega - k_x v_{i0})^2} - \frac{e^2 k^2 n_{b0}\varphi_1}{m_b\varepsilon_0(\omega - k_x v_{b0})^2} \tag{85}$$

On taking the plasma frequencies as; $\omega_{pe} = \sqrt{\frac{e^2 n_{eo}}{m_e \varepsilon_0}}$, $\omega_{pi} = \sqrt{\frac{e^2 n_{io}}{m_i \varepsilon_0}}$, and $\omega_{pb} = \sqrt{\frac{e^2 n_{bo}}{m_b \varepsilon_0}}$. Then the above equation reduces in the form as

$$-k^2 \varphi_1 = \frac{\omega_{pe}^2 \hat{\omega} k^2 \varphi_1}{\Omega^2 (\omega - k_y v_{e0})} - \frac{\omega_{pi}^2 k^2 \varphi_1}{(\omega - k_x v_{i0})^2} - \frac{\omega_{pb}^2 k^2 \varphi_1}{(\omega - k_x v_{b0})^2} \qquad (86)$$

Since the perturbed potential is not zero i.e. $\varphi_1 \neq 0$ then we get

$$\frac{\omega_{pe}^2 \hat{\omega}}{\Omega^2 (\omega - k_y v_{e0})} - \frac{\omega_{pi}^2}{(\omega - k_x v_{i0})^2} - \frac{\omega_{pb}^2}{(\omega - k_x v_{b0})^2} + 1 = 0 \qquad (87)$$

This is the modified dispersion relation for the lower-hybrid waves under the effects of collisions and electrons beam density.

15. Analytical solutions under the limitations

Consider now waves propagating along the $\hat{y}$ direction, so that $k_x = 0$, which, in real thruster geometry, corresponds to azimuthally propagating, waves. We discuss below its limiting cases through Litvak and Fisch [78].

$$\omega << |k_y v_{e0}|, \qquad (88)$$

The solutions for the dispersion relation (57) can be obtained as follows:

$$\omega^2 \approx \frac{\left(\omega_{pi}^2 + \omega_{pb}^2\right) \Omega^2}{\left(\Omega^2 + \omega_{pe}^2\right) \left[1 + \frac{i \nu_e \omega_{pe}^2}{(\Omega^2 + \omega_{pe}^2) k_y v_{e0}}\right]} \qquad (89)$$

Since the last terms in the second square brackets of the denominator in the right-hand side of (89) are small, we obtain the following

$$\omega \approx \pm \sqrt{\frac{\Omega^2 \left(\omega_{pi}^2 + \omega_{pb}^2\right)}{\left(\Omega^2 + \omega_{pe}^2\right)}} \left[1 - \frac{i \nu_e \omega_{pe}^2}{2 k_y v_{e0} \left(\Omega^2 + \omega_{pe}^2\right)}\right] \qquad (90)$$

Finally, the growth rate γ of the resistive instability is calculated from (90) as follow

$$\gamma \approx \frac{\nu_e \omega_{pe}^2}{2 k_y v_{e0} \left(\Omega^2 + \omega_{pe}^2\right)} \times \sqrt{\frac{\Omega^2 \left(\omega_{pi}^2 + \omega_{pb}^2\right)}{\left(\Omega^2 + \omega_{pe}^2\right)}} \qquad (91)$$

The corresponding real frequency $\omega_r (\omega \equiv \omega_r \pm i\gamma)$ is obtained as

$$\omega_r \approx \sqrt{\frac{\Omega^2 \left(\omega_{pi}^2 + \omega_{pb}^2\right)}{\left(\Omega^2 + \omega_{pe}^2\right)}} \qquad (92)$$

The Eqs. (91) show that the growth of the high frequency instability depends on collision frequency, electron density, ion density, beam density, azimuthal wave

Parameters	Range
Magnetic field	$B_{0g} \sim 100 - 200G$
Axial Wave number	$Kx \sim 200 - 600/m$
Azimuthal Wave number	$Ky \sim 400 - 1200/m$
Collisional frequency	$v \sim 10^6/s$
Initial drift of electron	$u_0 \sim 10^6 m/s$
Initial drift of ions	$v_0 \sim 2 \times 10^4 - 5 \times 10^4 m/s$
Plasma density	$n_{e0} \sim 10^{18}, n_{i0} \sim 10^{18}, n_{b0} \sim 10^{17}/m^3$
Thruster channel diameter	$D \sim 4 - 10$ cm

Table 2.
Plasma parameters.

number, initial drift and on the applied magnetic field. On the other hand, the real frequency of the wave depends only on the magnetic field, electron plasma density, ion density and beam density. By tuning these parameters one can control the frequency of the generating wave. In the below **Table 2**, the different parameters of a Hall thruster are given [51–56].

16. Results and discussion

The Eqs. (91) and (92) are solved with MATLAB by using appropriate parameters given in **Table 2**. We plot various figures for investigating the variation of growth rate and real frequency of the instability with magnetic field B_0 and density of beam n_b, initial drift, collision frequency v and wave number. For these sets of parameters, only one dominated mode of the dispersion relation is plotted in the figures. **Figure 3** shows the variation of growth rate and real frequency for different values of magnetic field. The reason for the enhanced growth rate as well as real frequency can be understood based on Lorentz force and the electron collisions. Since the electrons have their drift in the y-direction, they experience the Lorentz force due to the magnetic field in the negative of x-direction, i.e., in the direction opposite to the ions drift. The higher Lorentz force helps these transverse oscillations to grow relatively at a faster rate owing to an enhancement in the frequency. On the other hand, this is quite plausible that larger cyclotron frequency of the electrons leads to stronger effects of the collisions because of which the resistive coupling becomes more significant and hence the wave grows at its higher rate. Opposite effect of the magnetic field was observed by Alcock and Keen in case of a drift dissipative instability that occurred in afterglow plasma [76]. Similarly studied are also investigated by Sing and Malik in magnetized plasma [51–56].

In **Figure 4**, we have plotted the variation of growth rate γ and real frequency with the azimuthal wavenumber in order to examine the growth of these waves, when the oscillations are of smaller or relatively longer wavelengths. Here, the oscillations of larger wave numbers (or smaller wavelengths) are found to have lower growth. The faster decay that is observed on the larger side of k is probably due to the stronger Landau damping. The growth rate shows parabolic nature but the real frequency is almost increases linearly with respect to azimuthal wave number. It means that oscillations of smaller wavelengths are most unstable. Kapulkin et al. have theoretically observed the growth rate of instability to directly

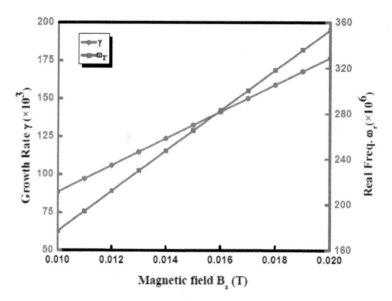

Figure 3.
Variation of growth rate and real frequency with the magnetic field.

proportional to the azimuthal wavenumber [77]. Litvak and Fisch have also shown that the rate of growth of instability is inversely proportional to the azimuthal wave number [78].

On the other hand, the variation of growth rate γ and real frequency with the collision frequency is depicted in **Figure 5**. The wave grow at faster rates in the presence of more electron collisions. This is due to the resistive coupling, which get much stronger in the presence of more collisions. In the present case, the growth rate grows at a much faster rate and real frequency is constant, and graph shows that the growth rate is directly proportional to the collision frequency. During the simulation studies of resistive instability, Fernandez *et al*. also observed the growth

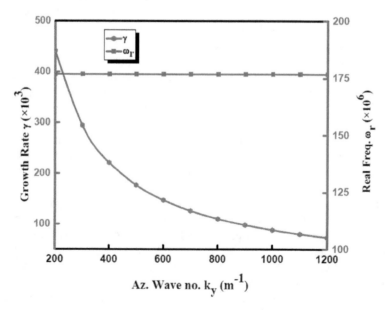

Figure 4.
Variation of growth rate γ and real frequency with azimuthal wavenumber.

rate to be directly proportional to the square root of the collision frequency [79]. In **Figure 6,** we show the dependence of the growth rate on the electron drift velocity. it is observed that the growth rate is reduced in the presence of larger electron drift velocity. In this case the resistive coupling of the oscillations to the electrons' drift would be weaker due to the enhanced velocity of the electrons. The reduced growth under the effect of stronger magnetic field is attributed to the weaker coupling of the oscillations to the electrons closed drift. The variation of growth rate γ and real frequency with beam density are shown in **Figure** 7. The growth shows asymmetric Gaussian type behavior but the real frequency varies linearly with beam density of electrons. This is due to the increased collisional effect with the large plasma density.

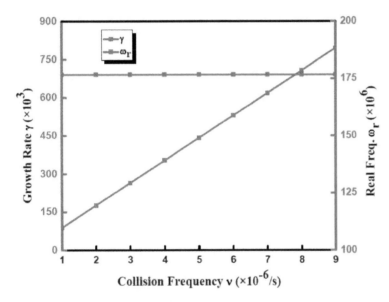

Figure 5.
Variation of growth rate γ and real frequency with collision frequency.

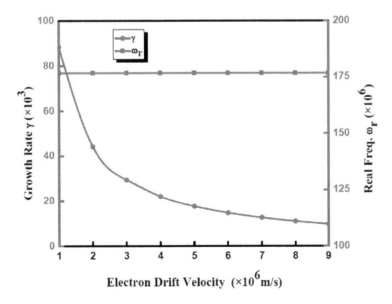

Figure 6.
Variation of growth rate γ and real frequency with electron drift velocity.

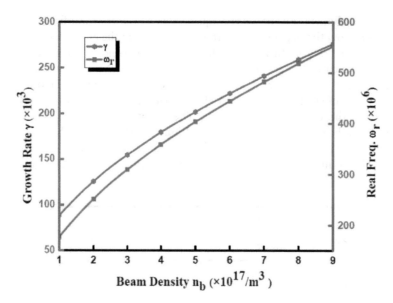

Figure 7.
Variation of growth rate γ and real frequency with beam density.

17. Conclusions

The present chapter discuss the properties of electromagnetic waves propagating through different media. In the first part of the chapter, the dispersion relation for the electromagnetic waves in conducting medium is derived. It has been experi-enced that the penetration of the electromagnetic field depend on the frequency of the source as well as the electrical properties of the medium. The significance of skin depth for biological and conducting media are explained through numerical examples. In the second part of the chapter, the generation of high frequency instability in plasma is discussed which grow with the magnetic field, wave length, collision frequency and the beam density. The growth rate linearly increases with collision frequency of electrons but it is decreases with the drift velocity of elec-trons. The real frequency of the instability increases with magnetic field, azimuthal wave number and beam density. The real frequency is almost independent with the collision frequency of the electrons.

Acknowledgements

The University Grants Commission (UGC), New Delhi, India is thankfully acknowledged for providing the startup Grant (No. F. 30-356/2017/BSR).

Author details

Sukhmander Singh[1*], Ashish Tyagi[2] and Bhavna Vidhani[3]

1 Plasma Waves and Electric Propulsion Laboratory, Department of Physics, Central University of Rajasthan, Ajmer, Kishangarh, India

2 Physics Department, Swami Shraddhanand College, University of Delhi, Delhi, India

3 Department of Physics and Electronics, Hansraj College, University of Delhi, Delhi, India

*Address all correspondence to: sukhmandersingh@curaj.ac.in

References

[1] L. Liu, S. Zhou and J. Cui, "Prospects and Problems of Wireless Communication for Underwater Sensor N e t-works," Wiley WCMC Special Issue on Underwater Sen-sor Networks (Invited), 2008.

[2] J. H. Goh, A. Shaw, A. I. Al-Shanmma'a, "Underwater Wireless Communication System," Journal of Physics, Conference Series 178, 2009.

[3] A. I. Al-Shamma'a, A. Shaw and S. saman, "Propagation of Electromagnetic Waves at MHz Frequencies through Seawater," IEEE Transactions on Antennas and Propa- tion, Vol. 52, No. 11, November 2004, pp. 2843-2849.

[4] S. Bogie, "Conduction and Magnetic Signaling in the Sea," Radio Electronic Engineering, Vol. 42, No. 10, 1972, pp. 447-452. doi:10.1049/ree.1972.0076

[5] Shan Jiang, Stavros Georgakopoulos, Electromagnetic Wave Propagation into Fresh Water, Journal of Electromagnetic Analysis and Applications, 2011, 3, 261-266

[6] Habib Mazharimousavi S, Roozbeh A, Halilsoy M. Electromagnetic wave propagation through inhomogeneous material layers. Journal of Electromagnetic Waves and Applications. 2013 Nov 1;27(16): 2065-74.

[7] Andrey B. Matsko and Yuri V. Rostovtsev, Electromagnetic-wave propagation and amplification in overdense plasmas: Application to free electron lasers, Physical Review E, 58, 6, DECEMBER 1998

[8] P. Sprangle, E. Esarey, J. Krall, and G. Joyce, Phys. Rev. Lett. 69, 2200 ~1992.

[9] .C. Oldano, Electromagnetic-wave propagation in anisotropic stratified media, Physical Review A, 40, 10 NOVEMBER 15, 1989

[10] Hidenori Akiyama, Takashi Sakugawa, Takao Namihira, Industrial Applications of Pulsed Power Technology, IEEE Transactions on Dielectrics and Electrical Insulation Vol. 14, No. 5; October 2007, 1051-1064

[11] A. Pokryvailo, M. Wolf, Y. Yankelevich, S. Wald, L. R. Grabowski, E. M. van Veldhuizen, Wijnand R. Rutgers, M. Reiser, B. Glocker, T. Eckhardt, P. Kempenaers and A. Welleman, "High-Power Pulsed Corona for Treatment of Pollutants in Heterogeneous Media", IEEE Trans. Plasma Sci., Vol. 34, pp. 1731-1743, 2006.

[12] G.J.J. Winands, K. Yan, A.J.M. Pemen, S.A. Nair, Z. Liu and E.J.M. vanHeesch, "An Industrial Streamer Corona Plasma System for Gas Cleaning", IEEE Trans. Plasma Sci., Vol. 34, pp. 2426-2433, 2006.

[13] W. Hartmann, T. Hammer, T. Kishimoto, M. Romheld and A. Safitri, "Ozone Generation in a Wire-Plate Pulsed Corona Plasma Reactor", 15th IEEE Pulsed Power Conf., pp. 856-859, 2005.

[14] J. Choi, T. Yamaguchi, K. Yamamoto, T. Namihira, T. Sakugawa, S. Katsuki and H. Akiyama, "Feasibility Studies of EMTP Simulation for the Design of the Pulsed-Power Generator Using MPC and BPFN for Water Treatments", IEEE Trans. Plasma Sci., Vol. 34, pp. 1744-1750, 2006.

[15] R.W. Bruce, R.L. Bruce, A.W. Fliflet, M. Kahn, S.H. Gold, A.K. Kinkead, D. Lewis, III and M.A. Imam, "Joining of ceramic tubes using a high-power 83-GHz Millimeter-wave beam", IEEE Trans. Plasma Sci., Vol. 33, pp. 668-678, 2005.

[16] A. M. Loske, "Medical and Biomedical Applications of Shock Waves: The State of the Art and the Near Future," in 30th International Symposium on Shock Waves 1, 2017, pp. 29-34.

[17] Claire M. Watts, Xianliang Liu, and Willie J. Padilla, Metamaterial Electromagnetic Wave Absorbers, Adv. Mater. 2012, 24, OP98–OP120, DOI: 10.1002/adma.201200674

[18] A. I. M. Ayala , Master of Science Thesis, Tufts University, USA, 2009 .

[19] Y. Gong , Z. Li , J. Fu , Y. Chen , G. Wang , H. Lu , L. Wang , X. Liu , Opt. Exp. 2011 , 19 , 10193

[20] Z. H. Jiang , S. Yun , F. Toor , D. H. Werner , T. S. Mayer , ACS Nano 2011 , 5 , 4641 .

[21] X. J. He , Y. Wang , J. M. Wang , T. L. Gui , PIER 2011 , 115 , 381

[22] E. Berry et al., Optical properties of tissue measured using terahertzpulsed imaging," Proc. SPIE, vol. 5030, pp. 459-470, Jun. 2003.

[23] A. J. Fitzgerald et al., Catalogue of human tissue optical properties at terahertz frequencies, J. Biol. Phys., vol. 29, nos. 2-3, pp. 123-128, 2003.

[24] G. W. Hanson, Dyadic Green's functions and guided surface waves for a surface conductivity model of graphene, J. Appl. Phys., vol. 103, no. 6, pp. 064302-1–064302-8, Mar. 2008.

[25] J. M. Jornet and I. F. Akyildiz, Graphene-based nano-antennas for electromagnetic nanocommunications in the terahertz band, in Proc. 4th Eur. Conf. Antennas Propag. (EUCAP), Apr. 2010, pp. 1-5.

[26] T. Binzoni, A.Vogel, A. H. Gandjbakhche, and R. Marchesini, Detection limits of multi-spectral optical imaging under the skin surface, Phys. Med. Biol., vol. 53, no. 3, pp. 617-636, 2008.

[27] Loske, A.M.: Medical and Biomedical Applications of Shock Waves. Shock Wave and High Pressure Phenomena. Springer International Publishing AG, Cham, Switzerland (2017). ISBN 978-3-319-47568-4

[28] Wang, C.J.: Extracorporeal shockwave therapy in musculoskeletal disorders. J. Orthop. Surg. Res. 7, 11–17 (2012)

[29] Kenmoku, T., Nobuyasu, O., Ohtori, S.: Degeneration and recovery of the neuromuscular junction after application of extracorporeal shock wave therapy. J. Orthop. Res. 30, 1660–1665 (2012)

[30] Delius, M., Hofschneider, P.H., Lauer, U., Messmer, K. Extracorporeal shock waves for gene therapy?Lancet 345, 1377 (1995)

[31] Chen ZY, Gao YM, Du M. Propagation characteristics of electromagnetic wave on multiple tissue interfaces in wireless deep implant communication. IET Microwaves, Antennas & Propagation. 2018 Jul 6;12 (13):2034-40.

[32] Callejon, M.A., Naranjo-Hernandez, D., Reina-Tosina, J. and Roa LM, Distributed circuit modeling of galvanic and capacitive coupling for intrabody communication', IEEE Trans. Biomed. Eng., 2012, 59, (11), pp. 3263–3269

[33] Seyedi M, Kibret B, Lai DT and Faulkner M. A survey on intrabody communications for body area network applications', IEEE Trans. Biomed. Eng., 2013, 60, (8), pp. 2067– 2079

[34] Marani R, Perri AG. RFID technology for biomedical applications: State of art and future developments. i-Manager's Journal on Electronics Engineering. 2015 Dec 1;6(2):1.

[35] A. Denisov and E. Yeatman, "Ultrasonic vs. inductive power delivery for miniature biomedical implants," in Body Sensor Networks (BSN), 2010 International Conference on, 2010, pp. 84-89: IEEE.

[36] S. Ozeri and D. Shmilovitz, "Ultrasonic transcutaneous energy transfer for powering implanted devices," Ultrasonics, vol. 50, no. 6, pp. 556-566, 2010.

[37] Li C, Un KF, Mak PI, Chen Y, Muñoz-Ferreras JM, Yang Z, Gómez-García R. Overview of recent development on wireless sensing circuits and systems for healthcare and biomedical applications. IEEE Journal on Emerging and Selected Topics in Circuits and Systems. 2018 Apr 3;8(2): 165-77.

[38] R. Muller et al., "A minimally invasive 64-channel wireless μECoG implant," IEEE Journal of Solid-State Circuits, vol. 50, no. 1, pp. 344-359, 2015.

[39] G. Papotto, F. Carrara, A. Finocchiaro, and G. Palmisano, "A 90nm CMOS 5Mb/s crystal-less RF transceiver for RF-powered WSN nodes," in Solid-State Circuits Conference Digest of Technical Papers (ISSCC), 2012 IEEE International, 2012, pp. 452-454: IEEE.

[40] Z.-P. Liang and P. C. Lauterbur, Principles of magnetic resonance imaging: a signal processing perspective. SPIE Optical Engineering Press, 2000.

[41] P. Lauterbur, "Image formation by induced local interactions: examples employing nuclear magnetic resonance," 1973.

[42] S. Gabriel, R. Lau, and C. Gabriel, "The dielectric properties of biological tissues: III. Parametric models for the dielectric spectrum of tissues," Physics in medicine and biology, vol. 41, no. 11, p. 2271, 1996.

[43] C. Gabriel, S. Gabriel, and E. Corthout, "The dielectric properties of biological tissues: I. Literature survey," Physics in medicine and biology, vol. 41, no. 11, p. 2231, 1996.

[44] S. Gabriel, R. Lau, and C. Gabriel, "The dielectric properties of biological tissues: II. Measurements in the frequency range 10 Hz to 20 GHz," Physics in medicine and biology, vol. 41, no. 11, p. 2251, 1996.

[45] Johnson C C and Guy AW, Nonionizing Electromagnetic Wave Effects in Biological Materials and Systems, Proceedings of THE IEEB, VOL. 60, NO. 6, J ~ T E(19 72)692-718

[46] C. J. Restall, P. F. Leonard, H. F. Taswell, and R. E. Holaday. IMPI Symp. (Univ. of Alberta, Edmonton, Canada, May 21-"Warming of human blood by use of microwaves," in Summ. 4th, 23, 1969), pp. 9699.

[47] Inan US, Said RK and Inan AS. Engineering electromagnetics and waves. Pearson; 2014 Mar 14.

[48] David J. Griffiths Introduction to Electrodynamics. (Addison-Wesley: Upper Saddle River, 1999).

[49] Jackson, J. D. (1998). Classical Electrodynamics. New York: Wiley, 3rd edition

[50] Mallick C, Bandyopadhyay M, Kumar R. Evolution of Microwave Electric Field on Power Coupling to Plasma during Ignition Phase. InSelected Topics in Plasma Physics 2020 Jun 15. IntechOpen.

[51] Tyagi J, Singh S, Malik HK, Effect of dust on tilted electrostatic resistive instability in a Hall thruster. Journal of Theoretical and Applied Physics. 2018; 12: 39-43. Doi.org/10.1007/s40094-018- 0278-z

[52] Singh S, Malik H K, Nishida Y. High frequency electromagnetic resistive

instability in a Hall thruster under the effect of ionization. Physics of Plasmas.2013; 20: 102109 (1-7).

[53] Singh S, Malik H K. Growth of low frequency electrostatic and electromagnetic instabilities in a Hall thruster. IEEE Transactions on Plasma Science.2011; 39:1910-1918.

[54] Singh S, Malik H K. Resistive instabilities in a Hall thruster under the presence of collisions and thermal motion of electrons. The Open Plasma Physics Journal. 2011; 4:16-23.

[55] Malik H K and Singh S. Resistive instability in a Hall plasma discharge under ionization effect. Physics of Plasmas.2013; 20: 052115 (1-8).

[56] Singh S. Evolutions of Growing Waves in Complex Plasma Medium. In edited book Engineering Fluid Mechanics. IntechOpen, London, United Kingdom, Nov 2020

[57] Singh S. Waves and Instabilities in E X B Dusty Plasma. In the edited book Thermophysical Properties of Complex Materials. IntechOpen, London, United Kingdom, December 12th 2019

[58] Singh S. Dynamics of Rayleigh-Taylor Instability in Plasma Fluids. In the edited book Engineering Fluid Mechanics. IntechOpen, London, United Kingdom, April 15th 2020

[59] Singh S. Hall Thruster: An Electric Propulsion through Plasmas. In the edited book Plasma Science IntechOpen, London, United Kingdom, March 2nd 2020 Doi.org/10.1063/1.2823033

[60] Singh S, Kumar S, Sanjeev, Meena S K and Saini S K. Introduction to Plasma Based Propulsion System: Hall Thrusters. In the edited book Propulsion - New Perspectives and Applications" edited by Prof. Kazuo Matsuuchi, IntechOpen, London, United Kingdom, March 2021

[61] Singh S, editor. Selected Topics in Plasma Physics. BoD–Books on Demand; 2020 Nov 19.

[62] Khalatpour A, Paulsen AK, Deimert C, Wasilewski ZR, Hu Q. High-power portable terahertz laser systems. Nature Photonics. 2021 Jan;15(1):16-20.

[63] Wu S, Cui S. Overview of High-Power Pulsed Power Supply. InPulsed Alternators Technologies and Application 2021 (pp. 1-35). Springer, Singapore.

[64] Wu S, Cui S. Electromagnetic Weapon Load of Pulsed Power Supply. InPulsed Alternators Technologies and Application 2021 (pp. 209-227). Springer, Singapore.

[65] Radasky, W. A., C. E., Baum, Wik, M. W.: Introduction to the special issue on high power electromagnetics (HPEM) and intentional electromagnetic interference (IEMI) environments and test capabilities. IEEE Trans. Electromagn. Compat. 46 (2004)

[66] Sabath, F., Backstrom, M., Nordstrom, B., Serafin, D., Kaiser, A., Kerr, B., Nitsch, D.: Overview of four European high power microwave narrow band test facilities. IEEE Trans. Electromagn. Compat. 46, 329 (2004)

[67] Parfenov, Y. V., Zdoukhov, L. N., Radasky, W. A., Ianoz, M.: Conducted IEMI threats for commercial buildings. IEEE Trans. Electromagn. Compat. 46, 404 (2004)

[68] Lu X, Picard JF, Shapiro MA, Mastovsky I, Temkin RJ, Conde M, Power JG, Shao J, Wisniewski EE, Peng M, Ha G. Coherent high-power RF wakefield generation by electron bunch trains in a metamaterial structure. Applied Physics Letters. 2020 Jun 29;116 (26):264102.

[69] Zhang J, Zhang D, Fan Y, He J, Ge X, Zhang X, Ju J, Xun T. Progress in

narrowband high-power microwave sources. Physics of Plasmas. 2020 Jan 17; 27(1):010501.

[70] Frank JW. Electromagnetic fields, 5G and health: what about the precautionary principle?. J Epidemiol Community Health. 2021 Jun 1;75(6): 562-6.

[71] Wu S, Cui S. Basic Theories of Pulsed Alternators. InPulsed Alternators Technologies and Application 2021 (pp. 37-61). Springer, Singapore.

[72] Aamodt, R. E., Sloan, M. L.: Nonlinear interactions of positive and negative energy waves. Phys. Fluids 11, 2218 (1968)

[73] Wilhelmson, H., Stenflo, I., Engelmann, F.: Explosive instabilities in the well defined phase description. J. Math. Phys. 11, 1738 (1970)

[74] O.P. Malik, Sukhmander Singh, Hitendra K. Malik, A. Kumar. Low and high frequency instabilities in an explosion- generated-plasma and possibility of wave triplet. Journal of Theoretical and Applied Physics (2015) Vol. 9 Pgs.75 -80.

[75] O.P. Malik, Sukhmander Singh, Hitendra K. Malik, A. Kumar. High frequency instabilities in an explosion- generated-relativistic-plasma. Journal of Theoretical and Applied Physics (2015) Vol. 9, Pgs.105-110.

[76] Alcock M W & Keen B E, Phys Rev A 3, (1971) 1087.

[77] Kapulkin A, Kogan A & Guelman M, Acta Astronaut, 55 (2004) 109.

[78] Litvak A A & Fisch N J, Phys Plasmas, 8 (2001) 648.

[79] Fernandez E, Scharfe MK, Thomas CA, Gascon N & Cappelli MA, Phys Plasmas, 15, 012102 (2008).

11

The Density Functional Theory and Beyond

Mohamed Barhoumi

Abstract

Density Functional Theory is one of the most widely used methods in quantum calculations of the electronic structure of matter in both condensed matter physics and quantum chemistry. Despite the importance of the density functional theory to find the correlation-exchange energy, but this quantity remains inaccurate. So we have to go beyond DFT to correct this quantity. In this framework, the random phase approximation has gained importance far beyond its initial field of application, condensed matter physics, materials science, and quantum chemistry. RPA is an approach to accurately calculate the electron correlation energy.

Keywords: DFT, LDA, GGA, RPA, Schrödinger equation

1. Introduction

The study of the microscopic properties of a physical system in the condensed matter branch requires the solution of the Schrödinger equation. When the studied system is composed of a large number of interacting atoms, the analytical solution of the Schrödinger equation becomes impossible. However, certain numerical cal-culation methods provide access to a solution to this fundamental equation for increasingly large systems. The calculation methods, called ab-initio like the density functional theory (DFT), propose to solve the Schrödinger equation without adjustable parameters. The density functional theory quickly established itself as a relatively fast and reliable way to simulate electronic and structural properties for all of the elements of the periodic table ranging from molecules to crystals. In this chapter, we recall the principle of this theory which considers electron density as a fundamental variable and that all physical properties can be expressed as a function of it.

2. Schrödinger equation

It is a fundamental equation to be solved to describe the electronic structure of a system with several nuclei and electrons and for a non-relativistic quantum description of a molecular or crystalline system and which is written:

$$H\Psi = \left(-\sum_i^n \frac{\hbar^2\nabla_i^2}{2m} - \sum_I^N \frac{\hbar^2\nabla_I^2}{2M} - \sum_{i,I} \frac{Z_I e^2}{|\overrightarrow{r_i} - R_I|} + \sum_{i<j} \frac{e^2}{|\overrightarrow{r_i} - \overrightarrow{r_j}|} + \sum_{I<J} \frac{Z_I Z_J e^2}{|R_I - R_J|}\right)\Psi,$$

(1)

where H is the molecular Hamiltonian and Ψ is the wave function. It is therefore a question of seeking the solutions of this equation. We can write the Hamiltonian in the form:

$$H = T_e + T_n + V_{e-e} + V_{n-n} + V_{n-e}.$$

(2)

We give the definition for each term:

$T_e = -\sum_i^n \frac{\hbar^2\nabla_i^2}{2m}$: The kinetic energy of n electrons of mass m.

$T_n = -\sum_I^N \frac{\hbar^2\nabla_I^2}{2M}$: The kinetic energy of N nuclei of mass M.

$V_{e-e} = \sum_{i<j} \frac{e^2}{|\overrightarrow{r_i} - \overrightarrow{r_j}|}$: The electron–electron repulsive potential energy.

$V_{e-n} = -\sum_{i,I} \frac{Z_I e^2}{|\overrightarrow{r_i} - \overleftarrow{R_I}|}$: The attractive potential energy nucleus-electron.

$V_{n-n} = \sum_{I<J} \frac{Z_I Z_J e^2}{|R_I - \overleftarrow{R_J}|}$: The nucleus-nucleus repulsive potential energy.

For a system of N nuclei and n electrons, Schrödinger equation is too complex to be able to be solved analytically. The exact solution of this equation is only possible for the hydrogen atom and hydrogenoid systems. In order to simplify the solution of this equation, Max Born and Robert Oppenheimer [1] have proposed an approximation aiming to simplify it.

3. The Born-Oppenheimer approximation

We consider that we can decouple the movement of electrons from that of nuclei, by considering that their movement of nuclei is much slower than that of electrons: we consider them as fixed in the study of the movement of the electrons of the molecule. The inter-nuclear distances are then treated as parameters. It has an immediate computational consequence, called an adiabatic hypothesis. It is in fact the same approximation and since the Oppenheimer approximation is still used in quantum chemistry, during chemical reactions or molecular vibrations, we can consider according to the classical Born-Oppenheimer approximation that the dis-tribution of electrons (adapts) almost instantaneously, when from the relative motions of nuclei to the resulting Hamiltonian variation. This is due to the lower inertia of the electrons $M = 1800m_e$ then the electron wave function can therefore be calculated when we consider that the nuclei are immobile, from where

$$T_n = 0; V_{n-n} = constant,$$

(3)

and so the Hamiltonian becomes

$$H = T_e + V_{e-e} + V_{e-n} + V_{n-n}.$$

(4)

$$H = H_{ele} + V_{n-n},$$

(5)

with H_{ele}: electronic Hamiltonian which is equal to:

$$H_{ele} = T_e + V_{e-e} + V_{e-n}.$$

(6)

Therefore the Born Oppenheimer approximation gives us:

$$H = T_e + V_{e-e} + V_{e-n}. \tag{7}$$

We use another notation to simplify the calculations

$$H = T + V_{ext} + U. \tag{8}$$

$$T = T_e; U = V_{e-e} = V_H; V_{ext} = V_{e-n}. \tag{9}$$

The Born-Oppenheimer approximation results in the Eq. (7) which keeps a very complex form: it always involves a wave function with several electrons. This approximation significantly reduces the degree of complexity but also the new wave function of the system depends on N bodies while other additional approximations are required to be able to effectively solve this equation. The remainder of this chapter will deal with approximations allowing to arrive at a solution of this equa-tion within the framework of the density functional theory (DFT) and the random phase approximation (RPA).

4. Density Functional theory (DFT)

Density Functional Theory is one of the most widely used methods for calculating the electronic structure of matter in both condensed matter physics and quantum chemistry. The DFT has become, over the last decades, a theoretical tool which has taken a very important place among the methods used for the description and the analysis of the physical and chemical properties for the complex systems, particularly for the systems containing a large number electrons. DFT is a reformulation of the N-body quantum problem and as the name suggests, it is a theory that only uses electron density as the fundamental function instead of the wave function as is the case in the method by Hartree and Hartree-Fock. The principle within the framework of the DFT is to replace the function of the multielectronic wave with the electronic density as a base quantity for the calculations. The formalism of the DFT is based on the two theorems of P. Hohenberg and W. Kohn [2].

4.1 Hohenberg and Kohn theorems

Hohenberg-Kohn (HK) reformulated the Schrödinger equation no longer in terms of wave functions but employing electron density, which can be defined for an N-electron system by:

$$n = 2N \int dr_1 \int dr_2 \ldots \int dr_{n-1} \Psi^*(r_1, r_2, \ldots r_{n-1}, r) \Psi(r_1, r_2, \ldots r_{n-1}, r), \tag{10}$$

this equation depends only on the three position parameters r = (x, y, z), position vector of a given point in space. This approach is based on two theorems demonstrated by Hohenberg and Kohn.

> **Theorem 1**: For any system of interacting particles in an external potential $V_{ext}\left(\overleftarrow{r}\right)$, the potential $V_{ext}\left(\overleftarrow{r}\right)$ is only determined, except for an additive constant, by the electron density $n_0\left(\overleftarrow{r}\right)$ in its ground state.

The first HK Theorem can be demonstrated very simply by using reasoning by the absurd. Suppose there can be two different external potentials $V_{ext}^{(1)}$ and $V_{ext}^{(2)}$ associated with the ground state density $n\left(\overleftarrow{r}\right)$. These two potentials will lead to two different Hamiltonians $H^{(1)}$ and $H^{(2)}$ whose wave functions $\psi^{(1)}$ and $\psi^{(2)}$ describing the ground state are different. As described by the ground state of $H^{(1)}$ we can therefore write that:

$$E^{(1)} = \left\langle \psi^{(1)}|H^{(1)}|\psi^{(1)}\right\rangle < \left\langle \psi^{(2)}|H^{(1)}|\psi^{(2)}\right\rangle. \tag{11}$$

This strict inequality is valid if the ground state is not degenerate which is supposed in the case of the approach of HK. The last term of the preceding expres-sion can be written:

$$\left\langle \psi^{(2)}|H^{(1)}|\psi^{(2)}\right\rangle = \left\langle \psi^{(2)}|H^{(2)}|\psi^{(2)}\right\rangle + \left\langle \psi^{(2)}|H^{(1)} - H^{(2)}|\psi^{(2)}\right\rangle, \tag{12}$$

$$\left\langle \psi^{(2)}|H^{(1)}|\psi^{(2)}\right\rangle = E^{(2)} + \int\left[V_{ext}^{(1)}\left(\overleftarrow{r}\right) - V_{ext}^{(2)}\left(\overleftarrow{r}\right)\right]n_0\left(\overleftarrow{r}\right)d^3r, \tag{13}$$

$$E^{(1)} < E^{(2)} + \int\left[V_{ext}^{(1)}\left(\overleftarrow{r}\right) - V_{ext}^{(2)}\left(\overleftarrow{r}\right)\right]n_0 d^3r. \tag{14}$$

It will also the same reasoning can be achieved by considering $E^{(2)}$ instead of $E^{(1)}$. We then obtain the same equation as before, the symbols (1) and (2) being inverted:

$$E^{(2)} = \left\langle \psi^{(2)}|H^{(2)}|\psi^{(2)}\right\rangle < \left\langle \psi^{(1)}|H^{(2)}|\psi^{(1)}\right\rangle, \tag{15}$$

$$\left\langle \psi^{(1)}|H^{(2)}|\psi^{(1)}\right\rangle = \left\langle \psi^{(1)}|H^{(1)}|\psi^{(1)}\right\rangle + \left\langle \psi^{(1)}|H^{(2)} - H^{(1)}|\psi^{(1)}\right\rangle, \tag{16}$$

$$\left\langle \psi^{(1)}|H^{(2)}|\psi^{(1)}\right\rangle = E^{(1)} + \int\left[V_{ext}^{(2)}\left(\overleftarrow{r}\right) - V_{ext}^{(1)}\left(\overleftarrow{r}\right)\right]n_0\left(\overleftarrow{r}\right)d^3r. \tag{17}$$

$$E^{(2)} < E^{(1)} + \int\left[V_{ext}^{(2)} - V_{ext}^{(1)}\right]n_0 d^3r, \tag{18}$$

we obtain the following contradictory equality:

$$E^{(1)} + E^{(2)} < E^{(1)} + E^{(2)}. \tag{19}$$

The initial hypothesis is therefore false; there cannot exist two external potentials differing by more than one constant leading at the same density of a non-degenerate ground state. This completes the demonstration.

$\Rightarrow$ the external potential of the ground state is a density functional.

Theorem 2: The previous theorem only exposes the possibility of studying the system via density. It only allows knowledge of the density associated with the studied system. The Hohenberg-Kohn variational principle partially answers this problem:

a universal functional for the energy $E[n]$ can be defined in terms of the density. The exact ground state is the overall minimum value of this functional.

Since the fundamental energy of the system is uniquely determined by its density, then energy can be written as a density functional. By following reasoning similar to that of the first part we show that the minimum of the functional corresponds to the energy of the ground state, indeed, the total energy can be written:

$$E_{HK}[n] = \int n\left(\overleftarrow{r}\right) V_{ext}\left(\overleftarrow{r}\right) d^3r + F_{HK}[n], \tag{20}$$

$F[n]$ is a universal functional of n(r):

$$F_{HK}[n] = T[n] + U[n]. \tag{21}$$

And the number of particles:

$$N = \int n\left(\overleftarrow{r}\right) dr. \tag{22}$$

Thus, we see that by minimizing the energy of the system with respect to the density we will obtain the energy and the density of the ground state. Despite all the efforts made to evaluate this functional E[n], it is important to note that no exact functional is yet known.

4.2 Ansatz of Kohn-Sham

Since the kinetic energy of a gas of interacting electrons being unknown, in this sense, Walter Kohn and Lu Sham [3] (KS) proposed in 1965 an ansatz which consists in replacing the system of electrons in interaction, impossible to solve analytically, by a problem of independent electrons evolving in an external potential. In the case of a system without interaction, the functional E[n] is reduced to kinetic energy and the interest of the reformulation introduced by Kohn and Sham is that we can now define a monoelectronic Hamiltonian and write the equations monoelectronic Kohn-Sham. According to KS the energy is written in the following form:

$$E_{HK}[n] = T_s[n] + \int V_{ext}(r)n(r)d^3r + E_{hartree}[n] + E_{xc}[n], \tag{23}$$

with the functional:

$$F_{HK}[n] = E_{HK}[n] - \int V_{ext}\left(\overleftarrow{r}\right) n\left(\overleftarrow{r}\right) d^3r. \tag{24}$$

$$F_{HK}[n] = T_s[n] + E_c[n] + E_{hartree}[n] + E_x[n] = T_s[n] + E_{hartree}[n] + E_{xc}[n]. \tag{25}$$

$T_s[n]$: representing the kinetic energy of a fictitious gas of non-interacting electrons but of the same density is given by:

$$T_s[n] = \sum_i \int dr \Psi_i^*\left(\overleftarrow{r}\right) \frac{-\nabla^2}{2} \Psi_i\left(\overleftarrow{r}\right). \tag{26}$$

$$E_{hartree}[n] = \frac{e^2}{8\pi\varepsilon_0} \int\int \frac{n\left(\overleftarrow{r}\right)n\left(\overleftarrow{r'}\right)}{|\overleftarrow{r} - r'|} d^3r d^3r'. \tag{27}$$

$T_s[n]$: the kinetic energy without interaction.
$E_{xc}[n]$: the exchange-correlation energy.
$E_{hartree}[n]$: the electron–electron potential energy.

$$E_{xc}[n] = E_{HK}[n] - \int V_{ext}\left(\overleftarrow{r}\right) n\left(\overleftarrow{r}\right) d^3r - T_s[n] - E_{hartree}[n]. \qquad (28)$$

$$E_{xc}[n] = F[n] - T_s[n] - E_{hartree}[n]. \qquad (29)$$

Based on the second Hohenberg-Kohn theorem, which shows that the electron density of the ground state corresponds to the minimum of the total energy and on the condition of conservation of the number of particles

$$\delta N\left[n\left(\overleftarrow{r}\right)\right] = \int \delta n\left(\overleftarrow{r}\right) dr = 0, \qquad (30)$$

So we have:

$$\delta\left\{E_{HK}[n] - \mu\left(\int n\left(\overleftarrow{r}\right) d^3r - N\right)\right\} = 0. \qquad (31)$$

$$\frac{\delta E_{HK}[n]}{\delta n\left(\overleftarrow{r}\right)} = \mu, \qquad (32)$$

$$\frac{\delta T_s[n]}{\delta n\left(\overleftarrow{r}\right)} + v^{eff}\left(\overleftarrow{r}\right) = \mu, \qquad (33)$$

$$v^{eff}\left(\overleftarrow{r}\right) = V_{ext}\left(\overleftarrow{r}\right) + \frac{e^2}{8\pi\varepsilon_0} \int \frac{n\left(\overleftarrow{r'}\right)}{|\overleftarrow{r} - r'|} d^3r' + \frac{\delta E_{xc}[n]}{\delta n\left(\overleftarrow{r}\right)}, \qquad (34)$$

therefore, the kinetic energy without interaction $T_s[n]$ is determined by:

$$\frac{\delta T_s[n]}{\delta n\left(\overleftarrow{r}\right)} = \left(3\pi^2 n\right)^{\frac{5}{3}} \frac{\hbar^2}{2m} = \frac{\hbar^2}{2m} k_F^2. \qquad (35)$$

Finally, the mono-electronic Hamiltonian of Kohn-sham in atomic unit is put in the form:

$$H = \frac{-\nabla^2}{2} + v^{eff}. \qquad (36)$$

The Hamiltonian is iteratively computed, the self-consistency of a loop is reached when the variation of the calculated quantity is lower than the fixed convergence criterion. The wave functions are calculated by a conjugate gradient method (or equivalent). The density is built from the wave functions, convergence is reached when the density is sufficiently close to the density of the previous step. When seeking to optimize the atomic structure of the system, an additional loop is added. With each iteration of this loop, the atomic positions are changed. It is said that the system is minimized when the forces are lower than the convergence criterion on the amplitude of the forces.

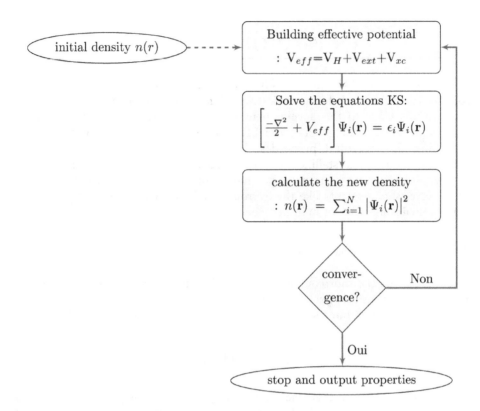

4.3 Expression of the exchange and correlation term

As described above, DFT is at the stage of Kohn-Sham equations, a perfectly correct theory insofar as the electron density which minimizes the total energy is exactly the density of the system of N interacting electrons. However, DFT remains inapplicable because the exchange-correlation potential remains unknown. It is therefore necessary to approximate this exchange-correlation potential. Two types of approximations exist the local density approximation or LDA and the generalized gradient approximation or GGA as well as the derived methods which are based on a non-local approach.

4.3.1 Local density approximation (LDA)

In only one model case, that of the uniform gas of electrons (corresponding quite well to the electrons of the conduction band of a metal), we know the exact expressions or with an excellent approximation of the terms of exchange and correlation respectively. In this LDA (Local Density Approximation), the electron density is assumed to be locally uniform and the exchange-correlation functional is of the form:

$$E_{xc}^{LDA} = \int n\left(\overleftarrow{r}\right) \varepsilon_{xc}^{homo}\left(n\left(\overleftarrow{r}\right)\right) dr. \tag{37}$$

$$\varepsilon_{xc}^{homo}\left(n\left(\overleftarrow{r}\right)\right) = \varepsilon_x[n] + \varepsilon_c[n]. \tag{38}$$

The function of $\varepsilon_{xc}^{homo}\left(n\left(\overleftarrow{r}\right)\right)$ is determined from a quantum computation parameterization for a constant electron density $n\left(\overleftarrow{r}\right) = n$;

$$\varepsilon_{xc}^{homo}\left(n\left(\overleftarrow{r}\right)\right) = \frac{3}{4}\left(\frac{3n(r)}{\pi}\right)^{\frac{1}{3}}. \tag{39}$$

Ceperley-Alder [4] numerically determined the contribution of the correlations. The search for analytical functions that come as close as possible to these results leads to the development of various functionalities with varying degrees of success. In general, the LDA approximation gives good results in describing the structural properties, i.e. it allows to determine the energy variations with the crystalline structure although it overestimates the cohesion energy, also concerning the mesh parameter for the majority of solids and good values of elastic constants like the isotropic modulus of compressibility. But this model remains insufficient in in-homogeneous systems.

4.3.2 Generalized gradient approximation (GGA)

To overcome the shortcomings of the LDA method, the generalized gradient approximation considers exchange-correlation functions depending not only on the density at each point, but also on its gradient [5], of the general form.

$$E_{xc}^{GGA}\left[n_\alpha, n_\beta\right] = \int n\left(\overleftarrow{r}\right)\varepsilon_{xc}\left[n_\alpha, n_\beta, \nabla_{n_\alpha(r)}\nabla_{n_\beta\left(\overleftarrow{r}\right)}\right]dr, \tag{40}$$

α and β are spins, in this case again, a large number of expressions have been proposed for this factor ε_{xc} leading to so many functionals. In general, the GGA improves compared to the LDA a certain number of properties such as the total energy or the energy of cohesion, but does not lead to a precise description of all the properties of a semiconductor material namely its electronic properties.

4.3.3 Functional hybrid HSE

The functions of DFT have been proved to be quite useful in explaining a wide range of molecular characteristics. The long-term nature of the exchange interaction, and the resulting huge processing needs, are a key disadvantage for periodic systems. This is especially true for metallic systems that necessitate BZ sampling. A new hybrid functionality, recently proposed by Heyd et al. [6], addresses this problem by separating the description of the exchange and the interaction into a short and long part. The expression of the exchange-correlation energy in HSE03 is given by:

$$E_{xc}^{HSE03} = \frac{1}{4}E_x^{sr,\mu} + \frac{3}{4}E_x^{PBE,sr,\mu} + E_x^{PBE,lr,\mu} + E_c^{PBE}. \tag{41}$$

As can be seen from the Eq. (41) only the exchange component of the electron–electron interaction is split into a short and long (lr) range (sr) part. The full electron correlation is represented by the standard correlation portion of the density of the GGA functional. Note that the term hybrid refers to the combined use of the exact exchange energy of the Hartree-Fock model and the exchange-correlation energy at the DFT level. The construction of hybrid functionals has been a good advancement in the field of exchange-correlation energy processing by allowing an explicit incorporation of the nonlocal character through the use of the exact term of exchange energy.

5. Random phase approximation

Despite the DFT is relevance in determining the exchange-correlation energy, it is still insufficient to characterize elastic characteristics. To rectify this amount, we must go beyond DFT. This is due to the random phase approximation as a method for calculating the electronic correlation energy accurately. RPA appears in the 1950s' [7–10] as a method of solving the N-body problem and arises from the desire to describe better (i.e. better than in a mean-field approximation) the physics of uniform electron gas, where the correlation between the positions of long-range electrons is important. In fact, collective oscillations (called plasma oscillations) are observed in an electron gas, which is the direct consequence of the long-range correlation between the electrons. Bohm and Pines, who introduced RPA, propose to place these collective oscillations at the center of solving the N-body problem, hoping that a good description of one will provide a good understanding of the other. RPA has been used with some success in the literature to describe systems containing Van der Waals interactions and in particular involving [11] scattering forces, which are known to be difficult to process. RPA introduced within the framework of DFT via the fluctuation-dissipation theorem with adiabatic connection (ACFDT).

6 Adiabatic-connection fluctuation-dissipation theory

The adiabatic connection fluctuation-dissipation (AC-FDT) [12] technique will be explained in order to discover the exact exchange-correlation energy in RPA. It will serve as the starting point for introducing the random phase approximation because it provides a general formulation for the exact correlation energy.

6.1 Adiabatic-connection (AC)

The adiabatic connection (AC) is a way to express the exact exchange-correlation energy function. The central idea in this approach is to build an interpolation Hamiltonian, which connects a Hamiltonian of an independent particle (reference Hamiltonian) $\hat{H}_0 = \hat{H}(\lambda = 0)$ and the physical Hamiltonians (multibody Hamiltonian) $\hat{H} = \hat{H}(\lambda = 1)$, with λ being a connection parameter. The AC technique can be used to derive the total energy of the ground state of a Hamiltonian of multiple interacting bodies, in which a continuous set of Hamiltonians dependent on the coupling force (λ) is introduced by:

$$\hat{H}(\lambda) = \hat{H}_0 + \lambda\hat{H}_1(\lambda) = \hat{T} + V_{ne} + \hat{V}(\lambda) + \lambda W_{ee}$$
$$= \sum_{i=1}^{N}\left[\frac{-1}{2}\nabla_i^2 + v_\lambda^{ext(i)}\right] + \sum_{i>j=1}^{N}\frac{\lambda}{|r_i - r_j|}. \quad (42)$$

With N being the number of electrons, v_λ^{ext} is an external potential with $v_{\lambda=1}^{ext}(r) = v^{ext}(r)$, being the external physical potential of the fully interactive system. Additionally, v_λ^{ext} can be spatially non-local for $\lambda \neq 1$. Following that, the reference Hamiltonian, or the Hamiltonian for an independent particle specified by the Eq. (42) for $\lambda = 0$, is of the mean field type, or is known in English as (Mean-field (MF)), i.e., a simple synthesis on a single-particle Hamiltonian:

$$\hat{H}_0 = \sum_{i=1}^{N}\left[\frac{-1}{2}\nabla_i^2 + v_{\lambda=0}^{ext}(r_i)\right] = \sum_{i=1}^{N}\left[\frac{-1}{2}\nabla_i^2 + v^{ext}(r_i) + v^{MF}(r_i)\right]. \quad (43)$$

With v^{MF} is an average field potential resulting from the electron–electron interaction. It can be the Hartree-Fock (HF) potential (v^{HF}) or the Hartree-plus correlation-exchange potential (v^{Hxc}) in the DFT. According to the two Eqs. (42) and (43), the perturbative Hamiltonian becomes:

$$\hat{H}_1(\lambda) = \sum_{i>j=1}^{N} \frac{1}{|r_i - r_j|} + \frac{1}{\lambda}\sum_{i=1}^{N}\left[v_\lambda^{ext}(r_i) - v_{\lambda=0}^{ext}(r_i)\right]$$
$$= \sum_{i>j=1}^{N} \frac{1}{|r_i - r_j|} + \frac{1}{\lambda}\sum_{i=1}^{N}\left[v_\lambda^{ext}(r_i) - v^{ext}(r_i) - v^{MF}(r_i)\right]. \qquad (44)$$

In the construction of the total energy, the ground state wave function $|\Psi_\lambda\rangle$ is introduced for the system λ, such that

$$H(\lambda)|\Psi_\lambda\rangle = E(\lambda)|\Psi_\lambda\rangle. \qquad (45)$$

Adopt the normalization condition, $\langle\Psi_\lambda|\Psi_\lambda\rangle = 1$, the total interacting energy of the ground state can then be obtained using the theorem of Hellmann-Feynman [13]

$$E(\lambda = 1) = E_0 + \int_0^1 d\lambda \times \left\langle\Psi_\lambda\right|\ \hat{H}_1(\lambda) + \lambda\frac{d\hat{H}_1(\lambda)}{d\lambda}\bigg)|\Psi_\lambda\right\rangle, \qquad (46)$$

The energy of order zero is E_0. It should be noted that the adiabatic connecting path chosen in Eq. (46) is not unique. In DFT, the path is chosen so that the electron density remains constant throughout the journey. This suggests a λ-dependency that $\hat{H}_1(\lambda)$ is not aware of.

6.2 The random phase approximation in the framework of adiabatic-connection fluctuation-dissipation theory

We will quickly discuss the concept of RPA in the context of DFT, which has served as the foundation for current RPA computations. The total ground state energy for an interacting N electron system is a (implicit) function of the electron density n(r) in the Kohn-Sham approximation (KS-DFT) and can be divided into four terms:

$$E[n(r)] = T_s[\psi(r)] + E_{ext}[n(r)] + E_H[n(r)] + E_{xc}[\psi_i(r)]. \qquad (47)$$

In the KS framework, the electron density is obtained from the single particle $\psi_i(r)$ orbitals via $n(r) = \sum_i^{occ}|\psi_i(r)|^2$. Among the four terms of the Eq. (47) only $E_{ext}[n(r)]$ and $E_H[n(r)]$ are explicit functions of n(r). T_s is treated exactly in KS-DFT in terms of single particle $\psi_i(r)$ orbitals which are themselves functional of n(r). The unknown correlation-exchange (XC) energy term, which is approximated as an explicit functionality of n(r) (and its local gradients) in conventional functional functions (LDA and GGA) and as a function of $\psi_i(r)$ in more advanced functions, contains the complete complexity of many bodies (hybrid density functions, RPA, etc.). In DFT, several existing approximations of E_{xc} can be categorized using a hierarchical approach called Jacob's scale [14]. But what if we wish to improve the accuracy of E_xc in a larger number of systems? To that purpose, starting with the technically accurate manner of generating E_xc using the AC technique mentioned above is instructive. As previously stated, the AC path is used in KS-DFT in order to maintain the correct electron density. Reducing the Eq. (46) for the total energy of the exact ground state E = E (λ = 1) to:

$$E = E_0 + \int_0^1 d\lambda \left\langle \Psi_\lambda | \frac{1}{2} \sum_{i \neq j=1}^N \frac{1}{|r_i - r_j|} | \Psi_\lambda \right\rangle + \int_0^1 d\lambda \left\langle \Psi_\lambda | \sum_{i=1}^N \frac{d}{d\lambda} v_\lambda^{ext}(r_i) | \Psi_\lambda \right\rangle$$

$$= E_0 + \frac{1}{2} \int_0^1 d\lambda \int\int drdr' \times \left\langle \Psi_\lambda | \hat{n}(r) \left[\frac{\hat{n}(r') - \delta(r-r')}{|r-r'|} \right] | \Psi_\lambda \right\rangle + \int drn(r) \left[v_\lambda^{ext}(r) - v_{\lambda=0}^{ext}(r) \right].$$

(48)

$$\hat{n}(r) = \sum_{i=1}^N \delta(r - r_i),$$

(49)

$\hat{n}(r)$ is the electron density operator and $n(r) = \langle \Psi_\lambda | \hat{n}(r) | \Psi_\lambda \rangle$, for any $0 \leqslant \lambda \leqslant 1$. For the reference state $|\Psi_0\rangle$ of KS (given by the slater determinant of orbitals $\Psi_i(r)$ occupied by a single particle, we get

$$E_0 = \left\langle \Psi_0 | \sum_{i=1}^N \left[-\frac{1}{2}\nabla^2 + v_{\lambda=0}^{ext}(r_i) \right] | \Psi_0 \right\rangle = T_s[\Psi_i(r)] + \int drn(r)v_{\lambda=0}^{ext}(r).$$

(50)

$$E = T_s[\Psi_i(r)] + \int drn(r)v_{\lambda=1}^{ext} + \frac{1}{2} \int_0^1 d\lambda \int\int drdr' \langle \Psi_\lambda | \frac{\hat{n}(r)[\hat{n}(r') - \delta(r-r')]}{|r-r'|} | \Psi_\lambda.$$

(51)

From the Eqs. (47) and (51), we obtained:

$$E_H[n(r)] = \frac{1}{2} \int drdr' \frac{n(r)n(r')}{|r-r'|}.$$

(52)

$$E_{ext}[n(r)] = \int drn(r)v_{\lambda=1}^{ext}(r).$$

(53)

We get the formally exact correlation-exchange energy expression XC;

$$E_{xc} = \frac{1}{2} \int d\lambda \int\int drdr' \frac{n_{xc}^\lambda(r,r')n(r)}{|r-r'|},$$

(54)

with $n_{xc}^\lambda(r,r')$ is defined by

$$n_{xc}^\lambda(r,r') = \frac{\langle \Psi_\lambda | \delta\hat{n}(r)\delta\hat{n}(r') | \Psi_\lambda \rangle}{n(r)} - \delta(r-r').$$

(55)

The mathematical expression for the so-called X C-hole is 55, with $\delta\hat{n} = \hat{n}(r) - n(r)$ denoting the fluctuation of the density operator $\hat{n}(r)$ around its expectation value n (r). The hole (XC) is also related to the density-density correlation function, as shown by the Eq. (55). It illustrates how the presence of an electron at point r reduces the density of all other electrons at point r' in physical terms. The temperature fluctuation-dissipation (FDT) theorem is used to relate the density-density correlations (fluctuations) in the Eq. (55) to the response (dissipation) features of the system in the second step. In statistical physics, FDT is a powerful approach. It shows that the reaction of a system in thermodynamic equilibrium to a tiny external disturbance is the same as the response to spontaneous internal fluc-tuations in the absence of disturbance [15]. FDT is applicable to both thermal and quantum mechanical fluctuations and shows itself in a variety of physical phenomena. A good example of the latter is the dielectric formulation of the many-body problem by Nozières and Pines [16]. The FDT at zero temperature performed at [16] is relevant in this situation.

$$\langle \Psi_\lambda | \delta \hat{n}(r) \delta \hat{n}(r') | \Psi_\lambda \rangle = -\frac{1}{\pi} \int_0^\infty d\omega \mathrm{Im} \chi^\lambda(r, r', \omega), \tag{56}$$

with $\chi^\lambda(r, r', \omega)$, is the linear density-response function of the system. Using the Eqs. (54) and (55) and $v(r, r') = \frac{1}{|rr'|}$, we arrive at the renamed ACFD expression for XC energy in DFT

$$E_{xc} = \frac{1}{2} \int_0^1 d\lambda \int\int dr dr' v(r, r') \times \left[-\frac{1}{\pi} \int_0^\infty d\omega \mathrm{Im} \chi^\lambda(r, r', \omega) - \delta(r - r') n(r) \right] = \frac{1}{2\pi} \int_0^1 d\lambda \int\int dr dr'$$

$$\times v(r, r') \times \left[-\frac{1}{\pi} \int_0^\infty d\omega^\lambda(r, r', i\omega) - \delta(r - r') n(r) \right]. \tag{57}$$

The analytical structure of $\chi^\lambda(r, r', \omega)$ and the fact that it becomes real on the imaginary axis are the reasons why the above frequency integration can be conducted along the imaginary axis. The problem of computing the energy XC on one of the response functions of a succession of fictional systems along the path AC is transformed by the expression ACFD in the Eq. (57), which must also be tackled in practice. RPA is a particularly basic approximation of the response function in this context:

$$\chi^\lambda_{RPA}(r, r', i\omega) = \chi^0(r, r', i\omega) + \int dr_1 dr_2 \chi^0(r, r_1, i\omega) \times \lambda v(r_1 - r_2) \chi^\lambda_{RPA}(r_2, r', \omega), \tag{58}$$

$\chi^0(r, r_1, i\omega)$, is the response function of independent particles of KS of the reference system $\lambda = 0$ and is known explicitly in terms of orbitals $\psi_i(r)$ single particle (KS), orbital energies ε_i and occupancy factors f_i:

$$\chi^0(r, r', i\omega) = \sum_{ij} \frac{\left(f_i - f_j \right) \psi_i^*(r) \psi_j(r) \psi_j^*(r') \psi_i(r')}{\varepsilon_i - \varepsilon_j - i\omega}. \tag{59}$$

From the Eqs. (57) and (58), the energy XC in RPA can be split into an exchange-exact (EX) and the correlation term RPA:

$$E_{xc}^{RPA} = E_x^{EX} + E_{RPA}^c. \tag{60}$$

$$E_x^{EX} = -\sum_{ij} f_i f_j \int\int dr dr' \psi_i^*(r) \psi_j(r) v(r, r') \psi_j^*(r') \psi_i(r'). \tag{61}$$

$$E_c^{RPA} = \frac{1}{2\pi} \int_0^\infty d\omega \mathrm{Tr} \left[\ln \left(1 - \chi^0(i\omega) v \right) + \chi^0(i\omega) v \right]. \tag{62}$$

7. Approximation of pseudo-potentials

The goal is to study the ground state of a system made up of nuclei, core electrons and valence electrons. The heart electrons are often closely linked to

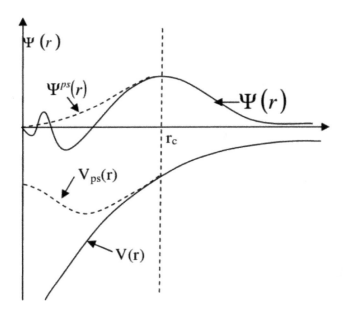

Figure 1.
Schematic illustration of all-electron potential (solid lines) and pseudo-potential (broken lines) and their corresponding wave functions.

nuclei, they are considered (frozen). This approximation makes it possible to develop the valence wave functions on a reduced number of plane waves having a kinetic energy lower than the energy of the cut-off (E_{cut} $E_{cut} \geqslant \frac{\hbar^2}{2m}|K+G|^2$), which allows correct treatment of the problem depends on the pseudo-potential used and the system studied. It consists in replacing the ionic potential $V_{el,nu}$ by a pseudopotential V^{ps} (see **Figure 1**) which acts on a set of wave pseudo-functions instead and places true wave functions and having the same eigenstates in the atomic Schrdinger equation. This idea has been developing since the end of the 1950s. This potential is constructed so as to reproduce the scattering properties for the true valence wave functions, while ensuring that the pseudo-wave function does not have a node in the core region defined by a cutoff radius r_c which is optimized for each orbital. Beyond the core region, the pseudopotential is reduced to the ionic potential so that the pseudo-wave function is equal to the true wave function. The use of a pseudo-potential reduces on the one hand the number of electrons consid-ered in the problem by taking into account only the valence electrons and on the other hand makes it possible to restrict the base of plane waves for the electrons of valence by eliminating most of the oscillations of the wave functions in the heart region.

8. Projection-augmented plane wave method (PAW)

We are always looking for the precision of the computation and thus to mini-mize the reasonable computation time as much as possible, we advise to use the least hard potentials possible, that is to say requiring the fewest plane waves to ensure convergence. The method of plane waves augmented by projection (PAW) (Projector Augmented Waves) [17] best meets this requirement, which explains their use in this thesis. The projection augmented plane wave method is an approach developed by P. Blochl, which models heart states from wave functions for an isolated atom; it assumes that these states are not modified when the atom is placed in a compound (approximation of frozen hearts). The PAW approach consists of using two kinds of basic functions: one uses partial atomic wave functions inside cores and the other uses a set of functions. Wave planes in the interstitial region. There is then a separation of space into two parts. The cause of this separation is the behavior of the effective potential seen by the valence shell electrons: in the inter-stitial zone, the total electron density is low. The wave functions oscillate rapidly which makes it possible to give strong variations in the amplitude of the total electron density. Because of these variations, the Fourier decomposition of the potential created by this charge distribution has components large wave vectors, which disadvantages the treatment in plane waves. We are therefore led to write in a different way the wave functions of the valence electrons in these two regions of space:

- In the interstitial region, plane waves are used.

- Inside the hearts, a partial wave decomposition solutions of the Schrödinger equation for the isolated atom which are much better adapted to the potential which prevails there.

9. Application

In spite of the significance of diamond and silicon, there's still a need for solid hypothetical and exploratory information on the elastic constants of these materials, in specific on the versatile constants of the third-order. This lack of experimental data limits the capacity of researchers to create modern materials with a focus on mechanical reactions. Besides, this lack triggered interest in other theoretical calculations. M. Barhoumi et al. [18] have proposed to calculate the elastic properties with different approximations of DFT and beyond with ACFDT in RPA, since the RPA has significant advantages, especially for those interested in functional density theory. It correctly describes the dispersion and van der Waals interactions. In this direction, they have found that the results obtained with RPA are in good agreement with the previous published [19–33]. Also, it should be noted that from the calculated elastic constants, other structural properties such as elastic modulus, shear modulus, Young's modulus and Poisson's ratio can be derived.

10. Conclusion

In this chapter, we have introduced the general method of calculating the ground state energy of a crystalline solid by application of DFT. We have just described how it is possible to determine the energy of the ground state of a solid by studying a fictitious system of independent particles giving rise to the same density as the real electronic system. On the other hand, we have highlighted the approxi-mations necessary to be able to apply this theory. Despite the importance of the DFT to find the exchange-correlation energy, but this quantity remains inaccurate to describe the elastic properties. So we have to go beyond DFT with RPA to correct this handicap. In this direction, we have shown that RPA is a good description of electronic correlation energy.

Author details

Mohamed Barhoumi
Faculté des Sciences de Monastir, Département de Physique, Laboratoire de la
Matière Condensée et des Nanosciences (LMCN), Université de Monastir,
Monastir, Tunisia

*Address all correspondence to: mohamedbarhoumi97@gmail.com

References

[1] M. Born et R. Oppenheimer. Annalen der Physik, 84, (1927) 457-484.

[2] P. Hohenberg. Phys. Rev. 136, (1964) B864-B864.

[3] W. Kohn et L. J. Sham, Phys. Rev. 50, (1965) A1133-A1138.

[4] D. M. Ceperley et B. J. Alder. Phys. Rev. Lett., 45, (1980) 566-569.

[5] J. P. Perdew, K. Burke et M. Ernzehof. Phys. Rev. Lett., 78, (1997) 385.

[6] HSE, J. Heyd, G. E. Scuseria, and M. Ernzerhof, J. Chem. Phys. 118, 8207 (2003)

[7] D. Bohm, and D. Pines. Phys. Rev. 82 (1951) 625.

[8] D. Pines, and D. Bohm. Phys. Rev. 85 (1952).

[9] D. Bohm, and D. Pines. Phys. Rev. 92 (1953) 609.

[10] M. Gell-Mann, and K. A. Brueckner. Phys. Rev. 106 (1957) 364.

[11] H. Rydberg, M Dion, N. Jacobson, E. Schroeder, P Hyldgaard, S. I. Simak, D. C. Langreth, and B. I. Lundqvist. Phys. Rev. Lett. 91 (2003).

[12] O. Gunnarsson et B. Lundqvist. Phys. Rev. B 13, (1976) 4274.

[13] H. Hellmann, Einfiihrung in die Quantenchemie (Deuticke,Leipzig,1937); R.P.Feynman, Phys. Rev. 56, (1939) 340.

[14] J. Perdew et K. Schmidt. In: Van Doren V, Van Alsenoy C, Geerlings P (eds). AIP, Melville, New York (2001).

[15] R. Kubo. Rep. Prog. Phys. 29, (1966) 255.

[16] D. P. Nozières, and D. Pines, A dielectric formulation of the many body problem: Application to the free electron gas. Nuovo Cim 9, 470-490 (1958).

[17] P. E. Blochl. Phys. Rev. B 50, (1994) 17953-17979.

[18] M.Barhoumi, D.Rocca, M.Said, and S.Lebgueb, Solid State Communications 324, (2021) 114136.

[19] A. Hmiel, J. M. Winey, and Y. M. Gupta, Phys. Rev B 93, (2016) 174113.

[20] Z. J. Fu, G. F. Ji, X. R. Chen, and Q. Q. Gou, Commun. Theor. Phys. 51, (2009) 1129.

[21] H. J. McSkimin, and J. P. Andreatch, J. Appl. Phys. 43, (1972) 2944.

[22] M. H. Grimsditch, and A. K. Ramdas, Phys. Rev. B 11, (1975) 3139.

[23] A. Migliori, H. Ledbetter, R. G. Leisure, C. Pantea, and J. B. Betts, J. Appl. Phys. 104, (2008) 053512.

[24] M. H. Grimsditch, E. Anastassakis, and M. Cardona, Phys. Rev. B 18, (1978) 901.

[25] E. Anastassakis, A. Cantarero, and M. Cardona, Phys. Rev. B 41, 7529 (1990).

[26] J. Winey, A. Hmiel, and Y. Gupta, J. Phys. Chem. Solids 93, (2016) 118.

[27] C. S. G. Cousins, Phys. Rev. B 67, (2003) 024107.

[28] O. H. Nielsen, and R. M. Martin, Phys. Rev. B 32, (1985) 3792.

[29] J. Zhao, J. M. Winey, and Y. M. Gupta Phys. Rev. B 75, (2007) 094105.

[30] M. Å opuzyski, and J. A. Majewski, Phys. Rev B 76, (2007) 045202.

[31] J. J. Hall, Phys. Rev. 161, (1967) 756-761.

[32] H. J. McSkimin, and P. Andreatch, J. Appl. Phys. 35, (1964) 3312.

[33] J. Philip, and M. Breazeale, J. Appl. Phys. 52, (1981) 3383.

12

Plasmonic 2D Materials: Overview, Advancements, Future Prospects and Functional Applications

Muhammad Aamir Iqbal, Maria Malik, Wajeehah Shahid, Waqas Ahmad, Kossi A. A. Min-Dianey and Phuong V. Pham

Abstract

Plasmonics is a technologically advanced term in condensed matter physics that describes surface plasmon resonance where surface plasmons are collective electron oscillations confined at the dielectric-metal interface and these collec-tive excitations exhibit profound plasmonic properties in conjunction with light interaction. Surface plasmons are based on nanomaterials and their structures; therefore, semiconductors, metals, and two-dimensional (2D) nanomaterials exhibit distinct plasmonic effects due to unique confinements. Recent technical breakthroughs in characterization and material manufacturing of two-dimen-sional ultra-thin materials have piqued the interest of the materials industry because of their extraordinary plasmonic enhanced characteristics. The 2D plasmonic materials have great potential for photonic and optoelectronic device applications owing to their ultra-thin and strong light-emission characteristics, such as; photovoltaics, transparent electrodes, and photodetectors. Also, the light-driven reactions of 2D plasmonic materials are environmentally benign and climate-friendly for future energy generations which makes them extremely appealing for energy applications. This chapter is aimed to cover recent advances in plasmonic 2D materials (graphene, graphene oxides, hexagonal boron nitride, pnictogens, MXenes, metal oxides, and non-metals) as well as their potential for applied applications, and is divided into several sections to elaborate recent theoretical and experimental developments along with poten-tial in photonics and energy storage industries.

Keywords: graphene, metal oxides, pnictogens, hBN, MXenes, non-metal plasmonics, photonics

1. Introduction

Plasmonics is the emerging research field, indicating the ability of materials to control light at nanoscale range to examine them for various properties and func-tions. The plasmonic materials exploit the surface plasmon resonance effects to achieve astonishing optical properties that originate with light-matter interaction and leads to remarkable results. Surface plasmon can confine electromagnetic fields at very small scales whereas various structures can be employed to control surface plasmons. Previously, Ag, Au, and Al metals were used as plasmonic materials but they did not perform well because of radiative losses, high amount of energy dissipation, and their poor tuneability. To overcome these problems for efficient plasmonic applications, a class of two-dimensional (2D) materials is proposed which presents a significant light-matter interaction phenomenon resulting in efficient quantum confinement effects. A variety of materials including semiconductors, conductive oxides, and dielectric materials have been investigated as plasmonic materials owing to their extra-ordinary plasmonic properties. Considering the advanced properties along with bandgap manipulation and electron transfer, 2D materials got higher attention for plasmonic applications [1, 2].

Graphene was the first 2D material investigated with zero bandgap having exceptional conductivity because of its high electron mobility. Considering graphene's achievements and enormous applications at the laboratory and industry level, researchers have started investigating further 2D materials to explore their potential for plasmonic applications. Currently, almost 150 members of the 2D materials family are serving in elementary and advanced technologies such as light-emitting diodes (LEDs), Field-effect transistors (FETs), environmental applications, sensing applications, and physical catalysis [3–5]. Some important under discussion members of 2D materials, analogous to graphene are; hexagonal boron nitride (hBN), black phosphorene, metal oxides, metal carbides and nitrides (MXenes), metal halides, pnictogens, and non-metals which are being considered as potential plasmonic materials [6–8]. This 2D materials family exhibits a broad electronic and plasmonic characteristic spectrum covering a wide range of proper-ties such as; high surface area, surface state nature, minimum dangling bonds, spin-orbit coupling, and quantum spin Hall effects [9, 10].

On the other hand, stacking of different 2D materials is also the emerging part of the material industry which yields novel heterostructure materials capable of introducing some building blocks in a materials family with enhanced physical and chemical properties. The novel 2D materials such as metal carbides and nitrides, metal oxides and graphene-based materials have mixed properties and can be further tuned by adjusting bandgap that would result in increased light-harvesting efficiency which is the basis to achieve desired optical, elec-tronic, and optoelectronic properties, making them promising materials for plas-monic applications [11, 12]. In addition, the plasmonic efficiency of 2D materials can also be enhanced by injecting plasmonic hot electrons to alter carrier inten-sity in 2D materials for higher photocatalysis output [13]. The recent extension of plasmonic materials from traditional metals to semiconductors to semi-metal graphene are identified as an ideal materials for surface plasmon resonance in plasmonic structures and their subsequent applications needed to be addressed accordingly. Moreover, the coupling effects between excitons and plasmons for 2D materials are the growing research interests that profound further studies for light-matter interactions to discover novel materials for innovative device applications.

2. Overview of 2D materials

The radiation-matter interaction is more prominent in 2D materials because of their thin sheet structures and significant quantum confinement effects that lead to enhanced electronic and optical properties. Owing to their advanced nature, 2D materials are advantageously evaluated for plasmonic characteristics, and multiple studies have been conducted for hBN to investigate plasmon molecular vibration coupling, plasmon substrate phonon coupling, and graphene plasmon-phonon polaritons coupling [14–16]. The 2D graphene structure exhibits exciting results due to its single-atom thickness and their environmental sensitivity. Other than environmental sensitivity, graphene plasmons can also be tuned with external mag-netic and electric fields [17]. The effectiveness of graphene-based plasmonics can be determined by charge carrier density, and heavily doped graphene exhibits high efficiency which is required for plasmonic applications [18]. As a result, graphene is an excellent plasmonic material, and combining graphene with other 2D materials is favorable to obtain optimum efficiency [19].

Hexagonal Boron Nitride (hBN) is one of the most intriguing findings in 2D materials for plasmonics, having the unique ability to be fabricated within the host material, and can be used as a promising substrate for graphene-based plasmonic applications because of its graphene-matched crystal structure [18, 20]. The hBN-graphene mixture is helpful to enhance grapheme-plasmon lifetime when compared with other 2D materials and can maintain its bandgap even in varied thicknesses depicting a wide range of plasmonic properties including electro-optic and quantum-optics [21, 22]. Moreover, the point defects in hBN at room tempera-ture demonstrate single-photon emission properties that can be used to integrate plasmonic nanostructures. Despite its wide bandgap, hBN offers high quantum efficiency, optical nonlinearity, and novel plasmonic properties to make it the best choice as 2D plasmonic material [23]. Its structure is shown in **Figure 1** [24].

The MXenes are a new class of 2D materials that contain carbides and nitrides, and they are the biggest family currently available, as seen in **Figure 2** [25]. This family of materials is substantially more stable than graphene with high metallic conductivity, folding and molding properties, and good electromagnetic properties, possessing the unique property of being combined with other materials to tune

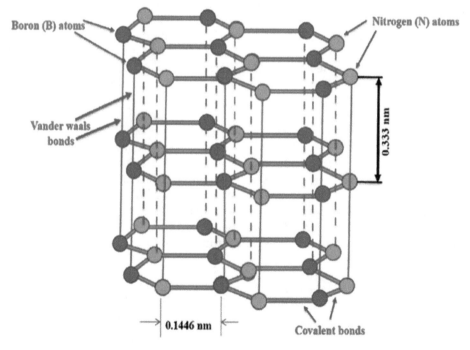

Figure 1.
Schematics of hBN structure [24].

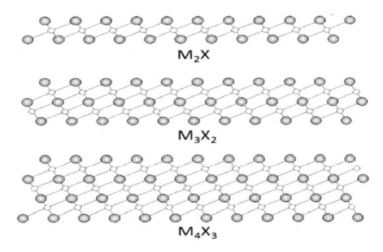

Figure 2.
MXene sheets with multiple layers where larger spheres represent transient metal, while smaller spheres are C, N, or CN [25].

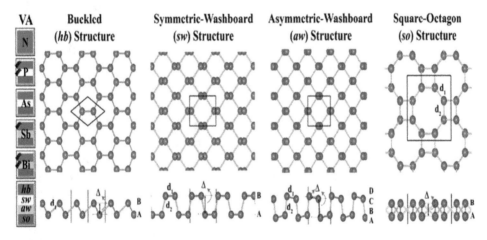

Figure 3.
Pnictogens and their schematic 2D structures [27].

their properties for desired applications. They can be employed in a variety of applications including energy storage devices, photonic-plasmonic structures as well as photocatalytic devices [26], and because of their metallic character along with high conductive nature, they may be used as plasmonic materials equivalent to graphene.

Researchers anticipated the group VA elements such as nitrogen, arsenic, antimony as well as bismuth as single-layer 2D structures with the introduction of 2D materials synthesis, and these elements are referred to as pnictogens and are shown in **Figure 3** [27]. These materials feature a honeycomb, washboard, and square-octagon structure, and they offer outstanding electrical, optical, electro-optical, and plasmonic properties having strong spin-orbit coupling, a narrow bandgap, and band inversion properties, making them ideal for plasmonic device applications [27].

3. Theoretical advancements

Photonics deals with the light-matter interaction which usually results in the formation of a single electron–hole pair by interacting light photons with free charge carriers in a metal, whereas in plasmonics, there is a large number of charge Carriers present that leads to collective oscillations which is the fundamental problem in plasmonics because all charge carriers are not part of the solid and can be influenced by structural defects as well as other materials defects such as disloca-tions. As a result, multi-scale modeling at various structural complexity levels is required for theoretical exploration of these complex models and plasmonic excita-tions in bulk materials and localized plasmons in metallic structures. To analyze this complicated issue, several theoretical and numerical models have been presented, although only a few of them are described here.

The Drude-Lorentz model which gives a theoretical insight into a material and can be employed in plasmonic applications is an intuitive way to study the underlying dielectric characteristics of solids [28, 29]. The Drude-Lorentz model, also known as the oscillator model, entails representing an electron as a driven damped harmonic oscillator in which the electron is connected to the nucleus by a hypothetical spring with an oscillating electric field acting as a driving force. It also describes the behavior of electrons in terms of their electro-optical characteristics when light interacts with them [30]. The Drude-Lorentz model's predictions are completely supported by the classical oscillator model as well as quantum mechanical features like electronic dipole moments of materials. To justify the microscopic qualities exhibited by classical and quantum techniques using this model, it is necessary to understand the Ehrenfest theorem which shows that quantum mechanical predicted values fundamentally follow classi-cal mechanical conditions [31]. The Kohn-Sham approach is praised for its ease of use in relating a many-body system to a non-interacting system and thereby solving it using Kohn-Sham density functional theory [32]. **Figure 4** illustrates such a model [32].

With the developments in computing, new algorithms are being devised to accomplish difficult jobs rapidly and accurately. Different approaches and models for electrical and photonic systems are being explored to accurately anticipate their characteristics, which were previously explored using differential equations. This section discusses the frequency domain approach and the time domain method to have a better knowledge of computational model advancements [33].

The decomposition of periodic systems into harmonic time-dependent eigen-modes is a basic approach for understanding the optoelectronic and plasmonic characteristics of materials. The frequency-domain approach is a subset of these decompositions that enlarges electromagnetic fields into Fourier eigenmodes which may be used to comprehend optical material properties in the absence of nonlinear effects [34]. This approach is usually started from fundamental photonic systems

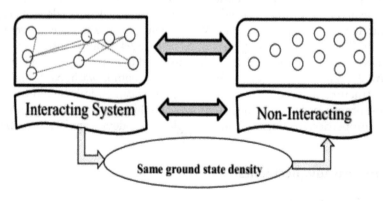

Figure 4.
Kohn-sham mapping of the interacting and non-interacting systems [32].

with translational symmetry which produces electromagnetic states and photonic band structures using Maxwell equations and wave Equations [35]. Although the frequency-domain technique is effective for defining material characteristics, but it is an expensive method that restricts its application in numerical models and as a result, the finite-difference time-domain method was presented as an alternative. The time-domain technique is a grid-based method that is linked to several other finite methods. This approach models electromagnetic wave propagation in dielectric media without needing derivation methods, making it easier to be utilized in complicated geometrical simulations, such as non-linear systems, which were previ-ously difficult to manage using the frequency-domain method. Furthermore, in this approach, Maxwell equations are discretized by differences arising from spatial and time derivatives, and the obtained results are solved in a leapfrog fashion on a staggered grid which is a good method being utilized in fluid dynamics [36]. Using the Phyton modules, these simulations can also be used to determine the plasmonic characteristics of materials [37].

The plasmonic material's behavior can be determined by studying the fre-quency-dependent dielectric factor linked with the excited state of the material. It is essential to analyze both the ground and excited states of material while calculating optical transitions based on material states. For the analysis of material character-istics based on these facts, *ab initio* methodologies like density functional theory (DFT), Hartree-Fock theory, and Green's function method [38–40] are essential. DFT [32] is founded on the concept that, for a quantum mechanical system, ground state charge density provides a complete comprehension of the system's ground state because the charge density of the state is mapped to the total energy of the system. The degree of freedom and functional complexity would be reduced if energy density is properly approximated. **Figure 5** [41] shows the bandgap differ-ence between graphene and hBN, which seems comparable but differs significantly at K-vector space as predicted within DFT (PBE) approximation computed by Warmbier et al. [41]. Green's function method is a quasi-particle approach for

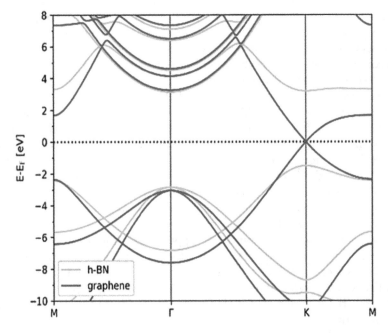

Figure 5.
Band structure of graphene and hBN computed within DFT (PBE) [41].

improving bandgap findings and reproducing band structures, with Hedin's GW as the most frequent implementation. All of these approaches hold promise for study-ing plasmonic material characteristics and predicting specific plasmonic device applications.

4. Experimental progress in the synthesis of 2D materials

The material characteristics such as its dimension, morphology, physical as well as chemical properties, orientation, and crystallinity mainly depend on the material's electronic properties which are impacted by the synthesis technique and experimental conditions. Mechanical exfoliation techniques have been tried in the past, but they have failed due to insufficient van-der-wall forces between 2D mate-rial layers, which limit uniformity and quality control, as well as their inability to scale-up [42]. Physical and chemical synthesis methods with controlled structural fabrication can be used with the top-down approach having the disadvantage of poor product yield and sheet restacking, limiting its application, and the bottom-up approach yielding promising results by assembling materials in a substrate using vapor deposition techniques such as physical vapor deposition (PVD), chemical vapor deposition (CVD), and atomic layer deposition. These are the most often utilized promising ways for fabricating 2D materials with customized thickness, dimensional control, high conductivity, and flexibility for electron transport; all of which are highly sought quantities for plasmonic applications [43].

CVD allows for the controlled synthesis of large areas of 2D materials with the added benefit of step-by-step film synthesis on various substrates and adjustable growth parameters to get the desired output. Metal–organic chemical vapor deposi-tion (MOCVD) is a modified version of CVD that is used to synthesize high-quality, large-area 2D materials for a variety of applications [44]. New research shows that a metal gas-phase precursor might be employed for regulated and uniform thick-ness instead of a powder precursor, which results in inhomogeneous nucleation and hence uncontrolled synthesis. Furthermore, temperature and pressure have an important impact on deposition uniformity, for example, high temperature and moderate pressure would result in excellent precursor coverage with regulated dimensions while excessively high temperatures might have negative consequences [45]. A comparison of various synthesis techniques is shown in **Figure 6** [46].

Many techniques are available for producing uniform 2D materials and hetero-structures with atomic layer deposition (ALD) being a refinement of the vapor-based deposition method in which the self-limiting reaction of the precursor is an essential aspect of ALD and the fact that a self-saturating surface monolayer is created after each precursor exposure distinguishes it from other deposition processes. In addition, ALD allows for the creation of 2D materials with fewer flaws and the synthesis of 2D material heterostructures with a small atom size thickness [47]. As a result of the advancements, ALD has opened up new means of synthesis with decreased interfacial impurities and large area deposition conformity with improved structural properties [48].

Apart from vapor-based synthesis approaches, liquid-phase exfoliation is another good way to get 2D materials in which the surface oxide of liquid metals produces unexpected results from a combination of physical and chemical features of liquid alloys [49]. While interacting with their ambient conditions, liquid metals with electron-rich metallic cores form a natural 2D film; the self-limiting surface oxide film with a thickness of a few atoms [50]. These liquid ingredients serve as host materials for the production of high-quality, one-kind films for innova-tive applications. The origin to define its development characteristics is the host

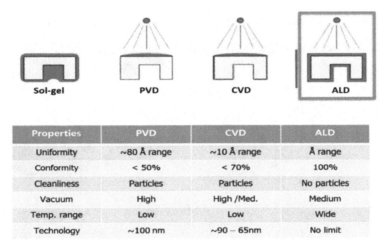

Figure 6.
Comparison of various 2D materials synthesis techniques [46].

materials fluidity, chemical composition, and thermodynamic properties that are the building block for examining the resultant 2D materials [51].

5. Plasmonic materials

Graphene, graphene oxides (GO), MXenes, pnictogens, and hBN are just a few of the commonly utilized plasmonic materials described in this section.

5.1 Graphene

Graphene is proving itself a revolutionary material for a wide range of applications since 2004 because of its electronic behavior which is responsible for exceptional features including, high mobility charge carriers, optical transmission, and tunable carrier densities [52–60]. The ability of graphene's structure to strongly confine excited surface plasmons in comparison to other materials as well as its ability to tune surface plasmons by manipulating charge densities is remarkable for prospective applications primarily in optoelectronics and plasmonics [61]. Experimental studies [62] show that graphene surface plasmons may be coupled with electrons and photons, allowing them to be used in more promising applications. This graphene coupling is in the form of quasi-particles that hold an intense interest in optoelectronics and condensed matter physics [63]. Surface plasmons in graphene offer a variety of significant advantages over other plasmonic materials, including high confinement, high tuneability, reduced frequency loss, improved electron relaxation time, and high many-body interactions. **Figure** 7 shows the structure and band description of graphene [64].

5.2 Graphene oxides (GO)

Graphene oxide (GO) is an amorphous insulator with carbon network bases on the hexagonal rings with both sp^2 and sp^3 hybridization as well as hydroxyl and epoxide groups on sheet sides and carboxyl and carbonyl groups on sheet edges. This kind of morphological structure is responsible for its wide range of technological applications mainly in nano-electronics, nano-photonics, and

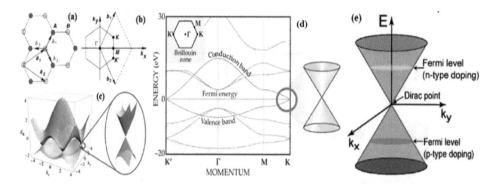

Figure 7.
(a) Graphene lattice structure, (b) BZ of graphene, (c) lattice electronic dispersion, (d) graphene electronic structure Bloch band description, (e) Low EBS approximation [64].

nano-composites. Because of the oxygen functionalities present, GO has a signifi-cant benefit when mixed with other polymeric or ceramic materials, resulting in improved electrical and mechanical properties [65]. The GO is in high demand in both industry and academics because of its zero-bandgap feature and excellent flexibility with superior thermal and electrical conductivity. In addition, one of GO's unique properties is the coexistence of its size, shape, and hybridization domains via a reduction mechanism that may eventually control the bandgap and convert GO to a semi-metal form. The main distinction between GO and reduced graphene oxide (r-GO) is that GO has oxygen-containing functional groups, whereas reduced graphene oxide does not [66]. Owing to heterogeneous electrical structure, GO is fluorescent throughout a wide wavelength range, while reduced GO allows quick response in GO-based electronics [67]. The GO is a promising choice for novel photonic materials, solar cells, optical devices, and a range of other applications due to its unique features. **Figure 8** depicts the difference between GO and r-GO [66].

5.3 Hexagonal boron nitride (hBN)

The hBN is a traditional 2D heterostructure material that was previously used as a substrate for thin-layered materials but now has the potential to be employed as an active plasmonic material. It is an excellent encapsulant for graphene because it protects it from the environment and increases its electrical mobility, extending the

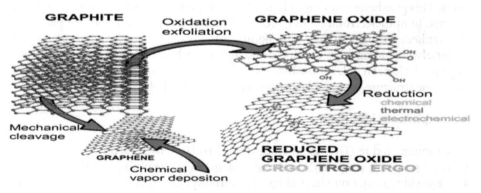

Figure 8.
Schematic flow from graphene-to-graphene oxide (GO) and reduced graphene oxide (r-GO) [66].

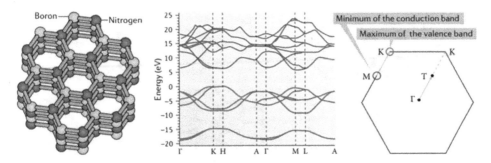

Figure 9.
hBN and its band structure properties [70].

life of surface plasmon polaritons in plasmonic applications [18, 68]. It has a crystal nature and polar bonding, allowing it to perform a wide range of optical, electrical, optoelectronic, and quantum optic functions for device applications. Also, it has a wide bandgap, high internal quantum efficiency, and significant optical nonlineari-ties that depend on material thickness and is specified by the rotation angle between heterostructure material layers [23, 69]. Its structure and band properties are displayed in **Figure 9** [70].

5.4 MXenes

MXenes are new types of 2D plasmonic materials made up of nitrides and carbides that were discovered in 2011. They are more stable than graphene, can be readily shaped and folded, have superior electromagnetic properties, and can be coupled with other materials to exhibit a wide range of applications in energy storage, supercapacitors, photonics, and plasmonics. The properties of MXenes can be determined by exploring surface termination, composition, doping, or mixing with other materials, resulting in adjustable conductivity that can change the material's properties as a metal or semiconductor [71]. The relative dielectric permittivity of MXenes for plasmonic applications can be studied by inter- and intra-band transitions that define optical parameters such as, absorption coef-ficient, refractive index, transmittance, and reflectance, and are linked to the material's electrical conductivity, which has already been demonstrated com-putationally and experimentally to find a place in electronic and optoelectronic applications [72, 73].

5.5 Pnictogens

Pnictogens are monolayer stable structures found from elements of group VA (nitrogen, arsenic, antimony, and bismuth) after the discovery of black phos-phorene. These materials are named as nitrogen in hexagonal buckled struc-ture, arsenic in hexagonal buckled as well as symmetric washboard structure, antimony, and bismuth in either hexagonal buckled or asymmetric washboard structure, but later on, these elements occurred to have other exotic struc-tures [74]. The pnictogens stability can be depicted using molecular dynamic simulations performed at high temperatures and materials phonon frequencies [75]. Pnictogens, as contrasted with group IV elements, are significantly more stable semiconductor materials with an appropriate bandgap for numerous device applications. Also, in contrast to black phosphorus (BP), they are ther-modynamically stable monolayer structures with rhombohedral structural

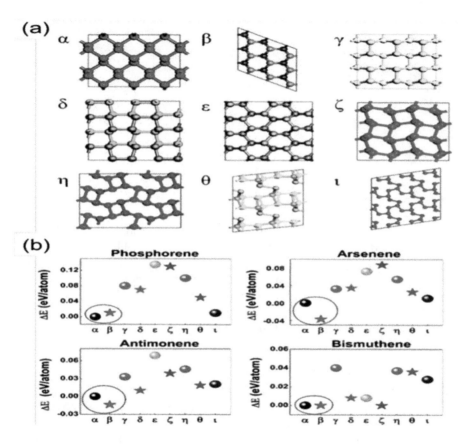

Figure 10.
(a) Honeycomb structures, (b) average binding energies of 2D pnictogens [77].

characteristics and interlayer covalent connections which decrease as anisotropy decreases and metallic character increases from arsenic to antimony to bismuth [76]. The 2D monolayer structures of pnictogens exhibit strong directionality in various physical properties that can be implied on plane lateral heterostructures to produce parallel strips of 2D pnictogens with advanced technological applica-tions while the same effect can be observed by a monoatomic chain of group VA elements attached to their monolayers. **Figure 10** represents structures as well as binding energies of pnictogens [77].

5.6 Metal oxides

Metal oxides show strong metallic behavior owing to stable charge carrier concentration when doped with various significant dopants such as aliovalent, oxygen vacancies, or interstitial dopants that result in localized surface plasmon resonance and by carefully choosing the host material as well as doped mate-rial, these surface plasmonic resonances can be tuned in the range of near- and mid-infrared (IR) region spectrum. The optical modeling of metal oxides illustrates the importance of defects and their impact on charge carrier mobility and the electronic structure of the material which reveals the choice of dopant as an important factor for metal oxides as plasmonic materials. Metal oxides are different from ordinary metals in the sense that they may change their localized surface plasmon resonance by changing their elemental composition, regardless of material size or shape, and these plasmon resonances can also be adjusted by altering external stimuli, resulting in the unique features of plasmonic materials as a result of crystal and morphological configurations that are useful for a variety of device applications [78].

6. Functional applications

The 2D materials can generally be categorized on the basis of electrical and optoelectronic properties in device applications such as flash memories, sen-sors, tunnel junctions, photodetectors, photonic crystals, optical metamaterials, nanophotonics, and quantum optics. The 2D graphene-based photonic and optoelectronic devices dragged much attention because of their versatile applica-tions in broad fields such as sensing, communication, and imaging technologies [79]. In terms of density of state and band structure, graphene has an adjustable light absorption spectrum and carrier density which may be used in waveguide-integrated graphene photonic devices and molecular sensor detection. Graphene with optical adjustments has also been used for light modulation and detection, and its derivatives are proving to be a feasible alternative for a variety of applica-tions. The GO can be utilized in the manufacture of electrical devices such as FETs and GFETs, LEDs, and solar cells, while r-GO dispersed in solvents may be utilized to replace FTO and ITO electrodes in transparent electrode manufac-ture; moreover, their large surface area and conductivity allow them to function as energy storage devices for longer periods with a greater capacity [80]. Recent investigations have shown that near-field IR optical microscopy and IR micros-copy of graphene are responsible for surface plasmon modeling in plasmonic applications [81]. An overview of 2D plasmonic materials-based devices is shown in **Figure 11** [82].

The hBN has exciting technological applications including, photonics, and its nanostructures feature weak polaritons that interact poorly with light and might be utilized to control the optical angular momentum of hyperbolic phonon polari-tons, implying maximum optical density of state and/or improving molecular IR vibrational absorption through surface enhancement [83]. Hybridization of hBN phonon-polariton with graphene surface plasmon-polaritons resulted in the active

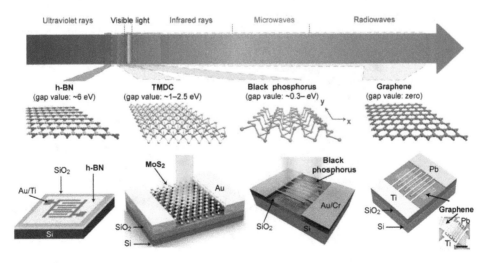

Figure 11.
2D plasmonic material-based devices for optoelectronic applications [82].

tuning of the polaritons which is a potential characteristic for chip-based nano-pho-tonics [84], photonic devices, modulators, and hyper-lensing [85]. The hBN defects that reduce phonon lifespan have uses in single-photon emitters (SPEs), which have certain appealing properties such as high quantum efficiency, optical stability, linear polarization, and high brightness. In addition, owing to its high efficiency and extended life in device applications, hBN has been utilized to replace AlGaN in deep-UV applications. Another method for incorporation of hBN is to link emission with plasmonic resonator-based structures in which localized surface plasmons cause broad field confinements throughout a wide range of emission, resulting in substantial Purcell amplification for dipole emission coupled to these resonators. When compared to uncoupled devices, the hBN quantum emitter coupling with plasmonic arrays has previously been demonstrated, with studies revealing PL enhancement and lifespan reduction with a quantum efficiency of around 40% and enhanced saturation count rates [86].

MXenes plasmonics is a relatively new field with a wide range of possible appli-cations, including surface-enhanced Raman spectroscopy, conductive substrates, and plasmonic sensing [72]. Nonlinear optical applications based on the nonlinear absorption process by plasmonic illumination near plasma frequency have been suggested for MXenes, and these nonlinear applications include ultrafast lasers, optical switching, and optical rectification devices like optical diodes. At near-infrared frequencies, arrays of two-dimensional titanium carbide ($Ti_3C_2T_x$) MXene nanodisks exhibit strongly localized surface plasmon resonances, which have been exploited to produce broadband plasmonic metamaterial absorber [87]. MXenes are also used as super-absorbers in broadband plasmonic metamaterials, and these super-absorbers may be used for photodetection and energy harvesting. **Figure 12** shows an MXenes super-absorber [88] with configurable nano-aperture width for broadband applications.

Pnictogens with a 2D structure are in great demand for high-performance device applications, since they have a midrange tunable bandgap and unparalleled mobility, allowing them to be employed in FETs for more efficient response than other materials. Because 2D materials lack a suitable bandgap, photodetectors are a major challenge, but pnictogens direct and tunable bandgap has solved this problem acting as a bridge between narrow and wider bandgap materials, attracting a lot of interest in photodetectors with improved photo-responsivity

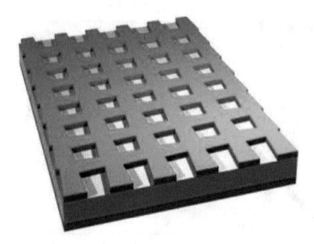

Figure 12.
MXene broadband super-absorber [88].

for telecommunication applications. The BP has an intriguing direct bandgap from visible to IR, making it a potential material for optoelectronic applications [89], while ultra-thin BP FETs have been described as effective NO_2 gas sensors with remarkable stability of pnictogens-based FETs during sensing [90, 91]. The 2D metals and metal oxide semiconductors, whether conducting or insulating, are useful for thin-film transistors and numerous device applications, where they can be employed in any component such as the source, drain, gate, electrodes, or gate dielectrics. Metal oxides are also useful in p-n junction device fabrications for diode rectifiers, solar cells, and organic photovoltaic applications, where they operate as a charge extracting interfacial layer to improve power conversion efficiency.

7. Future challenges and prospects

Plasmonics has advanced to the forefront of science due to technical advances in the experimental and computational fields as well as contributions to scien-tific applications. These contributions also confront some challenges that must be addressed in the future for effective plasmonic applications. To begin with, plasmonic nanostructures of controlled size and features cannot simply embed in their surroundings because they change the dielectric function of the surround-ing medium, affecting plasmonic switching and hence plasmonic applications. Another problem is optical pumping, which has the potential to deliver ultrafast plasmonic switching but has the drawbacks of destructive heat accumulation and high-power consumption. A major shortcoming of plasmonic materials is that self-tuned plasmonic structures lack effective plasmon coupling control abilities. Also, it is difficult to fabricate colloidal metal nano-crystals in controlled sym-metry for plasmonic device applications on a large scale, even though lithography techniques performed well but had some drawbacks such as high cost, long-time consumption, and difficulty with damped plasmonic properties on a large scale [92].

Plasmonics must control light at the nanoscale with minimal losses, and to do so, light localization must be pushed to new heights without jeopardizing its propaga-tion nature. Similarly, advances in topological plasmons must be incorporated in nanophotonic circuits by maintaining plasmon propagation stability and improving manufacturing techniques. For the experimental process to be effective, theoretical models must be improved to acquire the nonlinear and nonlocal physics of plas-monic devices. In short, both light and matter quantization are required to make a fine path toward a better understanding of light-matter interactions for advanced large-scale applications. The numerical approaches outlined are strong tools in terms of computing but they have conceptual limitations and their validity range becomes inefficient when a heterogeneous system is studied. To tackle plasmonic multi-scale challenges, the validity of numerical models must be improved by combining them with other numerical tools which is not well understood at this time and requires future considerations.

8. Conclusions

We have briefly addressed 2D plasmonic materials and their active properties in this chapter that are responsible for their wide range of applications in the electri-cal, photonic, and optoelectronic fields such as, FETs and GFETs, LEDs, and solar cells, modulators, hyper lensing, metamaterial absorbers, super-absorbers as well as nonlinear applications including ultrafast lasers, optical switching, and optical rectification devices like optical diodes. The synthesis techniques employed for 2D plasmonic materials have also been reviewed, with pulsed laser deposition (PLD) and CVD being the most extensively used and promising approaches for more con-trolled and conformational film growth. Also, these techniques have the advantage to provide desirable results by tuning their functional parameters such as tempera-ture, pressure, substrate angles, and deposition time. Computational models have to be examined to carry out a successful experiment, and there is a need to update simulation approaches to address problems in achieving desired plasmonic device features. Finally, we have outlined new prospective applications of 2D plasmonic materials and their significance in the industry as well as the drawbacks of materials that prohibit them from performing properly while providing the possible direc-tions for future research.

Conflict of interest

There are no conflicts of interest to declare.

Note

US spelling with serial comma.

Author details

Muhammad Aamir Iqbal[1], Maria Malik[2], Wajeehah Shahid[3], Waqas Ahmad[1,4], Kossi A. A. Min-Dianey[5,6] and Phuong V. Pham[6*]

1 School of Materials Science and Engineering, Zhejiang University, Hangzhou, China

2 Centre of Excellence in Solid State Physics, University of the Punjab, Lahore, Pakistan

3 Department of Physics, The University of Lahore, Lahore, Pakistan

4 Institute of Advanced Materials, Bahauddin Zakariya University, Multan, Pakistan

5 Département de Physique, Faculté Des Sciences (FDS), Université de Lomé, Lomé, Togo

6 Hangzhou Global Science and Technology Innovation Center, School of Micro-Nano Electronics, College of Information Science and Electronic Engineering, and Zhejiang University-University of Illinois at Urbana-Champaign Joint Institute (ZJU-UIUC), Zhejiang University, Hangzhou, China

*Address all correspondence to: phuongpham@zju.edu.cn; maamir@zju.edu.cn

References

[1] Geim AK, Novoselov KS. The rise of graphene. In: Nanoscience and Technology: A Collection of Reviews from Nature Journals. Singapore: World Scientific; 2010. pp. 11-19

[2] Novoselov KS, Morozov SV, Mohinddin TMG, Ponomarenko LA, Elias DC, Yang R, et al. Electronic properties of graphene. Physica Status Solidi. 2007;**244**(11):4106-4111

[3] Late DJ, Huang YK, Liu B, Acharya J, Shirodkar SN, Luo J, et al. Sensing behavior of atomically thin-layered MoS_2 transistors. ACS Nano. 2013;**7**(6):4879-4891

[4] Zhang Y, Chang TR, Zhou B, Cui YT, Yan H, Liu Z, et al. Direct observation of the transition from indirect to direct bandgap in atomically thin epitaxial $MoSe_2$. Nature Nanotechnology. 2014;**9**(2):111-115

[5] Pospischil A, Furchi MM, Mueller T. Solar-energy conversion and light emission in an atomic monolayer p–n diode. Nature Nanotechnology. 2014;**9**(4):257-261

[6] Nair RR, Blake P, Grigorenko AN, Novoselov KS, Booth TJ, Stauber T, et al. Fine structure constant defines visual transparency of graphene. Science. 2008;**320**(5881):1308-1308

[7] Wrachtrup J. Single photons at room temperature. Nature Nanotechnology. 2016;**11**(1):7-8

[8] Liu Y, Weiss NO, Duan X, Cheng HC, Huang Y, Duan X. Van der Waals heterostructures and devices. Nature Reviews Materials. 2016;**1**(9):1-17

[9] Tang Q, Zhou Z. Graphene-analogous low-dimensional materials. Progress in Materials Science. 2013;**58**(8):1244-1315

[10] Butler SZ, Hollen SM, Cao L, Cui Y, Gupta JA, Gutiérrez HR, et al. Progress, challenges, and opportunities in two-dimensional materials beyond graphene. ACS Nano. 2013;**7**(4): 2898-2926

[11] Zu S, Bao Y, Fang Z. Planar plasmonic chiral nanostructures. Nanoscale. 2016;**8**(7):3900-3905

[12] Zhu H, Yi F, Cubukcu E. Plasmonic metamaterial absorber for broadband manipulation of mechanical resonances. Nature Photonics. 2016;**10**(11):709-714

[13] Cheng F, Johnson AD, Tsai Y, Su PH, Hu S, Ekerdt JG, et al. Enhanced photoluminescence of monolayer WS2 on Ag films and nanowire–WS₂–film composites. ACS Photonics. 2017;**4**(6): 1421-1430

[14] Yang X, Kong XT, Bai B, Li Z, Hu H, Qiu X, et al. Substrate phonon-mediated plasmon hybridization in coplanar graphene nanostructures for broadband plasmonic circuits. Small. 2015;**11**(5):591-596

[15] Rodrigo D, Limaj O, Janner D, Etezadi D, De Abajo FJG, Pruneri V, et al. Mid-infrared plasmonic biosensing with graphene. Science. 2015;**349**(6244):165-168

[16] Brar VW, Jang MS, Sherrott M, Kim S, Lopez JJ, Kim LB, et al. Hybrid surface-phonon-plasmon polariton modes in graphene/monolayer h-BN heterostructures. Nano Letters. 2014; **14**(7):3876-3880

[17] Yan H, Li Z, Li X, Zhu W, Avouris P, Xia F. Infrared spectroscopy of tunable Dirac terahertz magneto-plasmons in graphene. Nano Letters. 2012;**12**(7):3766-3771

[18] Woessner A, Lundeberg MB, Gao Y, Principi A, Alonso-González P, Carrega M, et al. Highly confined low-loss plasmons in graphene–boron

nitride heterostructures. Nature Materials. 2015;**14**(4):421-425

[19] Yeung KY, Chee J, Yoon H, Song Y, Kong J, Ham D. Far-infrared graphene plasmonic crystals for plasmonic band engineering. Nano Letters. 2014;**14**(5):2479-2484

[20] Caldwell JD, Novoselov KS. Mid-infrared nanophotonics. Nature Materials. 2015;**14**(4):364-366

[21] Dai S, Ma Q, Liu MK, Andersen T, Fei Z, Goldflam MD, et al. Graphene on hexagonal boron nitride as a tunable hyperbolic metamaterial. Nature Nanotechnology. 2015;**10**(8):682-686

[22] Tran TT, Bray K, Ford MJ, Toth M, Aharonovich I. Quantum emission from hexagonal boron nitride monolayers. Nature Nanotechnology. 2016;**11**(1):37-41

[23] Ni GX, Wang H, Wu JS, Fei Z, Goldflam MD, Keilmann F, et al. Plasmons in graphene Moiré superlattices. Nature Materials. 2015;**14**(12):1217-1222

[24] Kumar A, Malik G, Chandra R, Mulik RS. Bluish emission of economical phosphor h-BN nanoparticle fabricated via mixing annealing route using non-toxic precursor. Journal of Solid State Chemistry. 2020;**288**:121 430

[25] Jakšić Z, Obradov M, Jakšić O, Tanasković D, Radović DV. Reviewing MXenes for plasmonic applications: Beyond graphene. In: 2019 IEEE 31st International Conference on Micro-electronics (MIEL). Washington, DC, USA: IEEE; 2019. pp. 91-94

[26] Anasori B, Lukatskaya MR, Gogotsi Y. 2D metal carbides and nitrides (MXenes) for energy storage. Nature Reviews Materials. 2017;**2**(2):1-17

[27] Ersan F, Keçik D, Özçelik VO, Kadioglu Y, Aktürk OÜ, Durgun E, et al. Two-dimensional pnictogens: A review of recent progresses and future research directions. Applied Physics Reviews. 2019;**6**(2):021308

[28] Hulst HC, van de Hulst HC. Light Scattering by Small Particles. New York, USA: Courier Corporation; 1981

[29] Bohren CF, Huffman DR. Absorption and Scattering of Light by Small Particles. Weinhein, Germany: John Wiley and Sons; 2008

[30] Mohammed F, Warmbier R, Quandt A. Computational plasmonics: Theory and applications. In: Recent Trends in Computational Photonics. Cham: Springer; 2017. pp. 315-339

[31] Sen D, Das SK, Basu AN, Sengupta S. Significance of Ehrenfest theorem in quantum–classical relationship. Current Science. 2001;**80**(4):536-541

[32] Iqbal MA, Ashraf N, Shahid W, Afzal D, Idrees F, and Ahmad R. Fundamentals of Density Functional Theory: Recent Developments, Challenges and Future Horizons. UK: IntechOpen; 2021

[33] Taflove A. Review of the formulation and applications of the finite-difference time-domain method for numerical modeling of electromagnetic wave interactions with arbitrary structures. Wave Motion. 1988;**10**(6):547-582

[34] Johnson SG, Joannopoulos JD. Block-iterative frequency-domain methods for Maxwell's equations in a plane-wave basis. Optics Express. 2001;**8**(3):173-190

[35] Joannopoulos JD, Johnson SG, Winn JN, Meade RD. Photonic Crystals: Molding the Flow of Light. 2nd ed. Princeton, New Jersey: Princeton Univ; 2008

[36] Smith GD, Smith GD, Smith GDS. Numerical Solution of Partial

Differential Equations: Finite Difference Methods. New York, USA: Oxford University Press; 1985

[37] Mohammed F, Warmbier R, Quandt A. Computational plasmonics: Numerical techniques. In: Recent Trends in Computational Photonics. New York City, USA: Springer; 2017. pp. 341-368

[38] Hohenberg P, Kohn W. Inhomogeneous electron gas. Physical Review. 1964;**136**(3B):B864

[39] Becke AD. A new mixing of Hartree–Fock and local density-functional theories. The Journal of Chemical Physics. 1993;**98**(2):1372-1377

[40] Hedin L. New method for calculating the one-particle Green's function with application to the electron-gas problem. Physical Review. 1965;**139**(3A):A796

[41] Warmbier R, Mehay T, Quandt A. Computational plasmonics with applications to bulk and nanosized systems. In: Active Photonic Platforms X. Vol. 10721. San Diego, California, United States: International Society for Optics and Photonics; 2018. p. 107211V

[42] Wei Z, Hai Z, Akbari MK, Qi D, Xing K, Zhao Q, et al. Atomic layer deposition-developed two-dimensional α-MoO3 windows excellent hydrogen peroxide electrochemical sensing capabilities. Sensors and Actuators B: Chemical. 2018;**262**:334-344

[43] Xu H, Akbari MK, Zhuiykov S. 2D Semiconductor nanomaterials and heterostructures: Controlled synthesis and functional applications. Nanoscale Research Letters. 2021;**16**(1):1-38

[44] Kang K, Xie S, Huang L, Han Y, Huang PY, Mak KF, et al. High-mobility three-atom-thick semiconducting films with wafer-scale homogeneity.Nature. 2015;**520**(7549):656-660

[45] Yu J, Li J, Zhang W, Chang H. Synthesis of high quality two-dimensional materials via chemical vapor deposition. Chemical Science. 2015;**6**(12):6705-6716

[46] Xu H, Akbari MK, Kumar S, Verpoort F, Zhuiykov S. Atomic layer deposition–state-of-the-art approach to nanoscale hetero-interfacial engineering of chemical sensors electrodes: A review. Sensors and Actuators B: Chemical. 2021;**331**:129403

[47] Leskelä M, Ritala M. Atomic layer deposition chemistry: Recent developments and future challenges. Angewandte Chemie International Edition. 2003;**42**(45):5548-5554

[48] Brun N, Ungureanu S, Deleuze H, Backov R. Hybrid foams, colloids and beyond: From design to applications. Chemical Society Reviews. 2011;**40**(2): 771-788

[49] Daeneke T, Khoshmanesh K, Mahmood N, De Castro IA, Esrafilzadeh D, Barrow SJ, et al. Liquid metals: Fundamentals and applications in chemistry. Chemical Society Reviews. 2018;**47**(11):4073-4111

[50] A de Castro I, Chrimes AF, Zavabeti A, Berean KJ, Carey BJ, Zhuang J, et al. A gallium-based magnetocaloric liquid metal ferrofluid. Nano Letters. 2017;**17**(12):7831-7838

[51] Caturla MJ, Jiang JZ, Louis E, Molina JM. Some Issues in Liquid Metals Research. Switzerland: MDPI; 2015.

[52] Britnell L, Gorbachev RV, Jalil R, Belle BD, Schedin F, Katsnelson MI, et al. Electron tunneling through ultrathin boron nitride crystalline barriers. Nano Letters. 2012;**12**(3):1707-1710

[53] Pham VP, Jang H-S, Whang D, Choi J-Y. Direct growth of graphene on rigid and flexible substrates: Progress, applications, and challenges. Chemical Society Reviews. 2017;**46**: 6276-6300

[54] Pham VP, Kim KH, Jeon MH, Lee SH, Kim KN, Yeom GY. Low damage pre-doping on CVD graphene/Cu using a chlorine inductively coupled plasma. Carbon. 2015;**95**:664 671

[55] Pham VP, Kim KN, Jeon MH, Kim KS, Yeom GY. Cyclic chlorine trap-doping for transparent, conductive, thermally stable and damage-free graphene. Nanoscale. 2014;**6**:15310-1 5318

[56] Pham VP, Kim, Nguyen MT, Park JW, Kwak SS, et al. Chlorine-trapped CVD bilayer graphene for resistive pressure sensor with high detection limit and high sensitivity. 2D Materials. 2017;**4**:025049

[57] Pham PV. Hexagon flower quantum dot-like Cu pattern formation during low-pressure chemical vapor deposited graphene growth on a liquid Cu/W substrate. ACS Omega. 2018;**3**:8036-8041

[58] Pham PV, Mishra A, Yeom GY. The enhancement of Hall mobility and conductivity of CVD graphene through radical doping and vacuum annealing. RSC Advances. 2017;**7**:16104- 16108

[59] Pham PV. Cleaning of graphene surfaces by low-pressure air plasma. Royal Society Open Science. 2018;**6**:172395

[60] Pham VP et al. Low energy BCl_3 plasma doping of few-layer graphene. Science of Advanced Materials. 2016;**8**:884-890

[61] Fang Z, Wang Y, Liu Z, Schlather A, Ajayan PM, Koppens FH, et al. Plasmon-induced doping of graphene. ACS Nano. 2012;**6**(11):10222-10228

[62] Fei Z, Rodin AS, Andreev GO, Bao W, McLeod AS, Wagner M, et al. Gate-tuning of graphene plasmons revealed by infrared nano-imaging. Nature. 2012;**487**(7405):82-85

[63] Zayats AV, Smolyaninov II, Maradudin AA. Nano-optics of surface plasmon polaritons. Physics Reports. 2005;**408**(3-4):131-314

[64] Cui L, Wang J, Sun M. Graphene plasmon for optoelectronics. Reviews in Physics. 2021;**6**:100054

[65] Mokhtar MM, Abo-El-Enein SA, Hassaan MY, Morsy MS, Khalil MH. Mechanical performance, pore structure and micro-structural characteristics of graphene oxide nano platelets reinforced cement. Construction and Building Materials. 2017;**138**:333-339

[66] Rowley-Neale SJ, Randviir EP, Dena ASA, Banks CE. An overview of recent applications of reduced graphene oxide as a basis of electroanalytical sensing platforms. Applied Materials Today. 2018;**10**:218-226

[67] Cao Y, Yang H, Zhao Y, Zhang Y, Ren T, Jin B, et al. Fully suspended reduced graphene oxide photodetector with annealing temperature-depende-nt broad spectral binary photorespon-ses. ACS Photonics. 2017;**4**(11):2797- 2806

[68] Kretinin AV, Cao Y, Tu JS, Yu GL, Jalil R, Novoselov KS, et al. Electronic properties of graphene encapsulated with different two-dimensional atomic crystals. Nano Letters. 2014;**14**(6): 3270-3276

[69] Kim CJ, Brown L, Graham MW, Hovden R, Havener RW, McEuen PL, et al. Stacking order dependent second harmonic generation and topological defects in h-BN bilayers. Nano Letters. 2013;**13**(11):5660-5665

[70] McCreary A, Kazakova O, Jariwala D, Al Balushi ZY. 2D Materials. 2020;**8**(1):013001-013018

[71] Dillon AD, Ghidiu MJ, Krick AL, Griggs J, May SJ, Gogotsi Y, et al. Highly conductive optical quality solution-processed films of 2D titanium carbide.

Advanced Functional Materials. 2016;**26**(23):4162-4168

[72] Hantanasirisakul K, Gogotsi Y. Electronic and optical properties of 2D transition metal carbides and nitrides (MXenes). Advanced Materials. 2018; **30**(52):1804779

[73] Liu Z, Alshareef HN. MXenes for optoelectronic devices. Advanced Electronic Materials. 2021;**7**(9):2100295

[74] Ersan F, Aktürk E, Ciraci S. Stable single-layer structure of group-V elements. Physical Review B.2016;**94**(24): 245417

[75] Özcelik VO, Aktürk OÜ, Durgun E, Ciraci S. Prediction of a two-dimensional crystalline structure of nitrogen atoms. Physical Review B. 2015;**92**(12):125420

[76] Gusmão R, Sofer Z, Bouša D, Pumera M. Pnictogen (As, Sb, Bi) nanosheets for electrochemical applications are produced by shear exfoliation using kitchen blenders. Angewandte Chemie. 2017;**129**(46):14609-14614

[77] Yua X, Lianga W, Xinga C, Chena K, Chena J, Huangd W, et al. Rising 2D pnictogens for catalytic applications: Status and challenges. Journal of. 2018;**6**(12):4883-5230

[78] Agrawal A, Johns RW, Milliron DJ. Control of localized surface plasmon resonances in metal oxide nanocrystals. Annual Review of Materials Research. 2017;**47**:1-31

[79] Tong L, Huang X, Wang P, Ye L, Peng M, An L, et al. Stable mid-infrared polarization imaging based on quasi-2D tellurium at room temperature.Nature Communications. 2020;**11** (1):1-10

[80] Li F, Jiang X, Zhao J, Zhang S. Graphene oxide: A promising nano - material for energy and environmental applications. Nano Energy. 2015;**16**:488-515

[81] Low T, Avouris P. Graphene plasmonics for terahertz to mid-infrared applications. ACS Nano. 2014;**8**(2):1086-1101

[82] Wang H, Li S, Ai R, Huang H, Shao L, Wang J. Plasmonically enabled two-dimensional material-based optoelectronic devices. Nanoscale. 2020;**12**(15): 8095-8108

[83] Autore M, Li P, Dolado I, Alfaro-Mozaz FJ, Esteban R, Atxabal A, et al. Boron nitride nanoresonators for phonon-enhanced molecular vibrational spectroscopy at the strong coupling limit. Light: Science and Applications. 2018;**7**(4):17172-17172

[84] Iqbal MA, Ashraf N, Shahid W, Awais M, Durrani AK, Shahzad K, Ikram M. Nanophotonics: Fundamentals, Challenges, Future Prospects and Applied Applications. UK: IntechOpen; 2021

[85] Caldwell JD, Vurgaftman I, Tischler JG, Glembocki OJ, Owrutsky JC, Reinecke TL. Atomic-scale photonic hybrids for mid-infrared and terahertz nanophotonics. Nature Nanotechnology. 2016;**11** (1):9-15

[86] Nguyen M, Kim S, Tran TT, Xu ZQ, Kianinia M, Toth M, et al. Nanoassembly of quantum emitters in hexagonal boron nitride and gold nanospheres. Nanoscale. 2018;**10**(5):2267-2274

[87] Chaudhuri K, Alhabeb M, Wang Z, Shalaev VM, Gogotsi Y, Boltasseva A. Highly broadband absorber using plasmonic titanium carbide (MXene). ACS Photonics. 2018;**5**(3):1115-1122

[88] Aydin K, Ferry VE, Briggs RM, Atwater HA. Broadband polarization-independent resonant light absorption using ultrathin plasmonic super

absorbers. Nature Communications. 2011;**2**(1):1-7

[89] Liu H, Du Y, Deng Y, Peide DY. Semiconducting black phosphorus: Synthesis, transport properties and electronic applications. Chemical Society Reviews. 2015;**44**(9):2732-2743

[90] Chen C, Youngblood N, Peng R, Yoo D, Mohr DA, Johnson TW, et al. Three-dimensional integration of black phosphorus photodetector with silicon photonics and nanoplasmonics. Nano Letters. 2017;**17**(2):985-991

[91] Abbas AN, Liu B, Chen L, Ma Y, Cong S, Aroonyadet N, et al. Black phosphorus gas sensors. ACS Nano. 2015;**9**(5):5618-5624

[92] Shao L, Tao Y, Ruan Q, Wang J, Lin HQ. Comparison of the plasmonic performances between lithographically fabricated and chemically grown gold nanorods. Physical Chemistry Chemical Physics. 2015;**17**(16):10861-10870

Permissions

The contributors of this book come from diverse backgrounds, making this book a truly international effort. This book will bring forth new frontiers with its revolutionizing research information and detailed analysis of the nascent developments around the world.

We would like to thank all the contributing authors for lending their expertise to make the book truly unique. They have played a crucial role in the development of this book. Without their invaluable contributions this book wouldn't have been possible. They have made vital efforts to compile up to date information on the varied aspects of this subject to make this book a valuable addition to the collection of many professionals and students.

This book was conceptualized with the vision of imparting up-to-date information and advanced data in this field. To ensure the same, a matchless editorial board was set up. Every individual on the board went through rigorous rounds of assessment to prove their worth. After which they invested a large part of their time researching and compiling the most relevant data for our readers.

The editorial board has been involved in producing this book since its inception. They have spent rigorous hours researching and exploring the diverse topics which have resulted in the successful publishing of this book. They have passed on their knowledge of decades through this book. To expedite this challenging task, the publisher supported the team at every step. A small team of assistant editors was also appointed to further simplify the editing procedure and attain best results for the readers.

Apart from the editorial board, the designing team has also invested a significant amount of their time in understanding the subject and creating the most relevant covers. They scrutinized every image to scout for the most suitable representation of the subject and create an appropriate cover for the book.

The publishing team has been an ardent support to the editorial, designing and production team. Their endless efforts to recruit the best for this project, has resulted in the accomplishment of this book. They are a veteran in the field of academics and their pool of knowledge is as vast as their experience in printing. Their expertise and guidance has proved useful at every step. Their uncompromising quality standards have made this book an exceptional effort. Their encouragement from time to time has been an inspiration for everyone.

The publisher and the editorial board hope that this book will prove to be a valuable piece of knowledge for researchers, students, practitioners and scholars across the globe.

List of Contributors

Luca Fasolo and Angelo Greco
Politecnico di Torino, Torino, Italy
INRiM - Istituto Nazionale di Ricerca Metrologica,
Torino, Italy

Emanuele Enrico
INRiM - Istituto Nazionale di Ricerca Metrologica,
Torino, Italy

Mohsin Ganaie
Centre for Energy Studies, Indian Institute of
Technology Delhi, Hauz Khas, New Delhi, India

Mohammad Zulfequar
Department of Physics, Jamia Millia Islamia, New
Delhi, India

Palle Kiran
Department of Mathematics, Chaitanya Bharathi
Institute of Technology, Hyderabad, Telangana,
India

Wen-Ping Wu, Yun-Li Li and Zhennan Zhang
Department of Engineering Mechanics, School of
Civil Engineering, Wuhan University, Wuhan,
China
State Key Laboratory of Water Resources and
Hydropower Engineering Science, Wuhan
University, Wuhan, China

Xiaoyang Zheng and Xian Zhang
Stevens Institute of Technology, Hoboken, New
Jersey, United States

Anca Armăşelu
Faculty of Electrical Engineering and Computer
Science, Department of Electrical Engineering and
Applied Physics, Transilvania University of Brasov,
Brasov, Romania

**Daniel Likius, Ateeq Rahman and Veikko
Uahengo**
Department of Chemistry and Biochemistry, Faculty
of Science, University of Namibia, Windhoek,
Namibia

Elise Shilongo
Department of Physics, Faculty of Science,
University of Namibia, Windhoek, Namibia

Ljudmila Shchurova
Division of Solid State Physics, Lebedev Physical
Institute, Russian Academy of Science, Moscow,
Russia

Vladimir Namiot
Department of Microelectronics, Scobelthsyn
Research Institute of Nuclear Physics, Moscow
State University, Moscow, Russia

Gulsum Ersu, Yenal Gokpek and Serafettin Demic
Department of Material Science and Engineering,
Faculty of Engineering and Architecture, Izmir
Katip Celebi University, Izmir, Turkey

Mustafa Can
Department of Engineering Sciences, Faculty of
Engineering and Architecture, Izmir Katip Celebi
University, Izmir, Turkey

Ceylan Zafer
Ege University, Institute of Solar Energy, İzmir,
Turkey

Sukhmander Singh
Plasma Waves and Electric Propulsion Laboratory,
Department of Physics, Central University of
Rajasthan, Ajmer, Kishangarh, India

Ashish Tyagi
Physics Department, Swami Shraddhanand College,
University of Delhi, Delhi, India

Bhavna Vidhani
Department of Physics and Electronics, Hansraj
College, University of Delhi, Delhi, India

Mohamed Barhoumi
Faculté des Sciences de Monastir, Département de
Physique, Laboratoire de la Matière Condensée et
des Nanosciences (LMCN), Université de Monastir,
Monastir, Tunisia

Muhammad Aamir Iqbal
School of Materials Science and Engineering,
Zhejiang University, Hangzhou, China

Maria Malik
Centre of Excellence in Solid State Physics, University of the Punjab, Lahore, Pakistan

Wajeehah Shahid
Department of Physics, the University of Lahore, Lahore, Pakistan

Kossi A. A. Min-Dianey
Département de Physique, Faculté Des Sciences (FDS), Université de Lomé, Lomé, Togo
Hangzhou Global Science and Technology Innovation Center, School of Micro-Nano Electronics, College of Information Science and Electronic Engineering, and Zhejiang University-University of Illinois at Urbana-Champaign Joint Institute (ZJU-UIUC), Zhejiang University, Hangzhou, China

Waqas Ahmad
Institute of Advanced Materials, Bahauddin Zakariya University, Multan, Pakistan
School of Materials Science and Engineering, Zhejiang University, Hangzhou, China

Phuong V. Pham
Hangzhou Global Science and Technology Innovation Center, School of Micro-Nano Electronics, College of Information Science and Electronic Engineering, and Zhejiang University-University of Illinois at Urbana-Champaign Joint Institute (ZJU-UIUC), Zhejiang University, Hangzhou, China

Index